Intelligent Transport Systems

TRANSPORT ECONOMICS, MANAGEMENT AND POLICY

General Editor: Kenneth Button, *Professor of Public Policy, School of Public Policy, George Mason University, USA*

Transport is a critical input for economic development and for optimizing social and political interaction. Recent years have seen significant new developments in the way that transport is perceived by private industry and governments, and in the way academics look at it.

The aim of this series is to provide original material and up-to-date synthesis of the state of modern transport analysis. The coverage embraces all conventional modes of transport but also includes contributions from important related fields such as urban and regional planning and telecommunications where they interface with transport. The books draw from many disciplines and some cross disciplinary boundaries. They are concerned with economics, planning, sociology, geography, management science, psychology and public policy. They are intended to help improve the understanding of transport, the policy needs of the most economically advanced countries and the problems of resource-poor developing economies. The authors come from around the world and will represent some of the outstanding young scholars as well as established names.

Titles in the series include:

Air Transport Networks
Theory and Policy Implications
Kenneth Button and Roger Stough

Analytical Transport Economics
An International Perspective
Edited by Jacob B. Polak and Arnold Heertje

Intelligent Transport Systems
Cases and Policies
Edited by Roger R. Stough

Transport and Environment
In Search of Sustainable Solutions
Edited by Eran Feitelson and Erik T. Verhoef

Environmental Costs and Liberalization in European Air Transport
A Welfare Economic Analysis
Youdi Schipper

Reforming Transport Pricing in the European Union
A Modelling Approach
Edited by Bruno De Borger and Stef Proost

Intelligent Transport Systems

Cases and Policies

Edited by

Roger R. Stough

NOVA Endowed Professor of Public Policy and Director of the National Center for ITS Implementation Research, George Mason University, USA

TRANSPORT ECONOMICS, MANAGEMENT AND POLICY

Edward Elgar
Cheltenham, UK • Northampton, MA, USA

Published by
Edward Elgar Publishing Limited
Glensanda House
Montpellier Parade
Cheltenham
Glos GL50 1UA
UK

Edward Elgar Publishing, Inc.
136 West Street
Suite 202
Northampton
Massachusetts 01060
USA

A catalogue record for this book
is available from the British Library

Library of Congress Cataloguing in Publication Data

Intelligent transport systems : cases and policies / edited by Roger Stough.
     p. cm. — (Transport economics, management, and policy series)
   Includes bibliographical references and index.
   1. Intelligent Vehicle Highway Systems. I. Stough, Roger. II. Transport economics, management, and policy

TE228.3 .I575 2001
388.3'12—dc21

                        2001023135

ISBN 1 84064 447 8

Printed and bound in Great Britain by MPG Books Ltd, Bodmin, Cornwall

Contents

Figures

Tables

Contributors

Brien Benson   George Mason University, Fairfax, VA

William M. Bowen   Cleveland State University, Cleveland, OH

Kingsley E. Haynes   George Mason University, Fairfax, VA

Dingjian Jin   Dickinson College, Carlisle, PA

Hans Klein   Georgia Institute of Technology, Atlanta, GA

Gerard Maas   George Mason University, Fairfax, VA, and Leiden University, The Netherlands

Roberto Mazzoleni   Hofstra University, Hempstead, NY

Mark E. Maggio   George Mason University, Fairfax, VA Hofstra University, Hempstead, NY

Michelle Sager   George Mason University, Fairfax, VA

Laurie Schintler   George Mason University, Fairfax, VA

Hadi Shafie   George Mason University, Fairfax, VA

Roger R. Stough   George Mason University, Fairfax, VA

Acknowledgments

This volume would not have been possible without the financial and other support provided by the Federal Highway Administration under a Cooperative Agreement (Agreement number: DTFH61-93-X-0027) and by the US Department of Transportation, Research and Special Projects Administration, University Transportation Center (UTC) Program (Grant number: TRS98-G-0013). The editor and authors gratefully acknowledge this support.

As with all such works, many are due thanks, not the least of whom are the authors of the various chapters, who processed several rounds of comments and redrafting to bring the volume to a publishable level. Aleta Wilson is due particular consideration as she assumed the onerous task of putting the manuscripts into a form acceptable to the publisher. Finally, Mary Clark is due thanks for the variety of ways she helped to support the completion of the book in a timely manner.

1. Introduction

Roger R. Stough

INTRODUCTION

Metropolitan areas have experienced accelerated decentralization over the past decade or two.[1] This has resulted in metropolitan areas that are geographically larger than heretofore with some, such as the US National Capital Region, growing in two decades from 50 or so miles in diameter to nearly 100 miles with no apparent end to "sprawling" in sight (Stough, 1995; Garreau, 1991). Increased demand for mobility by a more affluent and urban population, unique American preferences that seek a strong melding of "town and country" (Ewing, 1994; Vance, 1990), continued lower land costs on the expanding periphery and policies that support fringe development (tax, depreciation and building codes, and policies that discourage reinvestment and the reuse of urban and suburban land) are some of the factors responsible for spread effects (see Gordon and Richardson, 1996; Ewing, 1994; Bourne, 1992; Levernier and Cushing, 1994; Linneman and Summers, 1991; Stanback, 1991; Champion, 1989). These forces are inducing increased demand for mobility and transportation and thus the need for more transport capacity. Yet, it is increasingly difficult to push through public decisions creating more traditional transportation infrastructure such as new road lanes.

The growing impasse for the provision of traditional transportation infrastructure has stimulated a search for substitutes. Technology in the form of Intelligent Transportation Systems (ITS), behavioral change (for example, adoption of policies that support telework) as well as land use management changes (such as smart or concentrated development) are some of the alternative demand management approaches that have been considered and in part adopted.

THE PROMISE OF INTELLIGENT TRANSPORT SYSTEMS

This book focuses on the evaluation of the ITS alternative. ITS are a complex of technologies that are derived from information and computer technology

(ICT) and applied to transport infrastructure and vehicles. Some of the better-known ITS components are Advanced Traffic/Transportation Management Systems (ATMS), Advanced Transportation Information Systems (ATIS), and Commercial Vehicle Operations (CVO). These have a variety of sub-components that include, for example, collision avoidance, variable message signs (VMS), electronic guidance systems (EGS), electronic toll collection (ETC) and integrated information systems[2] that support improved cross-jurisdiction provision of emergency medical, police and fire services, and information for travelers. More specifically some of the individual technologies are electronic sensors, wire and wireless communication devices, computer hardware and software, global positioning systems (GPS) and even geographic information systems (GIS). All of these ITS components and systems are together envisioned to increase significantly the productivity and capacity of existing surface transportation infrastructure. It is also expected that ITS will enhance the ability to achieve other transportation goals including intermodality, systems integration, improved safety and environmental quality.[3]

Despite the vision of ITS providing potential capacity improvements as high as 20 percent, it is in a relatively nascent stage of development and deployment. Existing little more than 10 years, this group of technologies has for the most part been deployed on a demonstration basis.[4] Only recently have ITS systems begun to be adopted more widely but still, in most metropolitan areas, there is little more than a vision (in some cases a plan) of a deployable ATMS or ATIS system.[5,6] As a consequence, knowledge about how and to what extent these systems are increasing capacity and improving productivity, safety, mobility and air quality are not well known or appreciated. Further, despite increasing efforts to evaluate ITS projects, no widely agreed upon methodology has been developed or adopted.

PURPOSE OF THE BOOK

ITS evaluation is the primary concern of this book. Thus, in Chapter 2 the focus is on the identification and examination of a set of methodological and technical challenges facing the evaluation of ITS systems and projects. There, Roger R. Stough, Mark Maggio and Dingjian Jin conclude from their assessment of ITS evaluation approaches that, at a general level, a multi-criterion approach is more appropriate than other methodologies such as cost-benefit, cost-effectiveness and financial analysis. This conclusion is reached because it is difficult to measure and even more difficult to monetize many of the concerns of non-traditional groups that are now affected by transportation decisions and investments. A modest multi-criterion assessment tool in the form of a checklist methodology and template is provided (see Appendix,

Chapter 2) and applied to specific ITS deployment projects in the subsequent chapters of the book. In sum, the purposes of the book are to examine the ITS evaluation question, to offer a generalized approach and methodology for ITS project evaluation and to demonstrate the application of the methodology as well as other methodologies through empirical case studies of ITS deployments.

EVALUATION CHALLENGES

There are two general types of impediments or challenges to the evaluation of ITS deployments. One is more methodological in nature and is concerned with the viability of the assumptions adopted in more conventional approaches to transport investment evaluation including:

- project time horizon;
- vision and vision development for the region where ITS is being implemented;
- potential increasing returns to investments;
- demand estimation;
- induced demand;
- induced vs. direct effects of investments;
- deployment context – regional vs. corridor vs. national;
- subjectivity;
- innovation and adoption effects;
- ability to measure (exactness).

The other set of challenges is more technical in nature and focuses on the data, models and techniques that have been used to conduct conventional transport evaluations. In Chapter 2, it is noted that these techniques have not been able to overcome a variety of shortcomings. The outcome dimensions where data, modeling and measurement problems exist are:

- safety;
- congestion–delay;
- capacity–mobility–network integration;
- construction and installation costs;
- economic (employment, income and output);
- fiscal;
- environmental;
- equity;
- institutional.

The proposed approach to evaluation of ITS projects is derived from an assessment of these methodological challenges and technical contexts where data, modeling and measurement issues are significant challenges. It is important to note that methodological and technical challenges are not always mutually exclusive. For the most part, areas of overlap have been managed in an effort to minimize this non-exclusivity.

The assessment of the evaluation process for ITS, as indicated above, assumes a need to address the expanding involvement of new communities of interest in transport planning and decision making (Stough and Rietveld, 1997). Increasing involvement of groups such as environmentalists, economic development proponents, those concerned with equity issues (for example, prejudice, ethnicity and poverty), and institutional issues (such as those concerned with privacy or specific property rights issues) has occurred. As a consequence, the approach to evaluation of ITS here assumes that an important aspect is the issue of whom the evaluation is for. Given the expansion of participants and therefore their standing in the transport planning and decision making process, a full cost-benefit and outcomes oriented approach was adopted and embedded in the general multi-criterion assessment framework. This is examined more fully in Chapter 2.

As noted above, ITS is a recent development. Thus, the case-study evaluations in this book are more *ex ante* than *ex post* oriented. This occurs because many of the projects considered were being implemented while the evaluations were being conducted. While the *ex ante* approach is dominant, some *ex post* evaluation components are included in nearly all of the case studies and in at least several of them (Chapters 4, 5 and 6) significant parts of the assessment are *ex post* in nature.

REGIONAL FOCUS OF INTELLIGENT TRANSPORT SYSTEMS AND EVALUATION OF THEIR DEPLOYMENT

Historically in the US, transportation investment and infrastructure maintenance have been largely the responsibility of either specific local jurisdictions or state government. This means that systems for managing transportation either have been highly decentralized among local political jurisdictions (for example, the State of Maryland) or highly centralized under the direction of a state-wide agency (for example, the Commonwealth of Virginia). Regardless of the model, deployment and management of transport infrastructure has not historically been focused at the cross-jurisdictional metropolitan regional or transport corridor levels. Yet a major potential benefit of ITS is that it often produces greater benefits when implemented at the region-wide level because the information needed to improve traffic flow is regional and thus must be collected and/or managed in a coordinated and integrated way across jurisdictions. This asymmetry between existing management systems and the need for systems to manage information and traffic across jurisdictions at the metropolitan or corridor levels is a major impediment to the deployment of ITS. This occurs because non-traditional institutional and organizational infrastructure in the focus of multi-jurisdictional organizations often does not exist or has limited authority to act, such as many of the Metropolitan Councils of Government. Case studies that form the subsequent chapters of this book are focused at the regional level for this reason.

Each of the case studies presented in Chapters 3–9 are taken from the US National Capital Region. This region is a good laboratory for such studies because there have been a number of tests as well as actual deployments of ITS in the region. Further, it is institutionally complex in that governance is for the most part divided among two states (Maryland and Virginia) and a federal district (Washington, DC). Finally, it is a large metropolitan region that has very high congestion levels (second highest in the US after Los Angeles). Thus it is a region that may benefit considerably from ITS deployment. Figure 1.1 is provided for the reader to better comprehend the region and the location of its jurisdictions and Interstate highways. This map is referred to

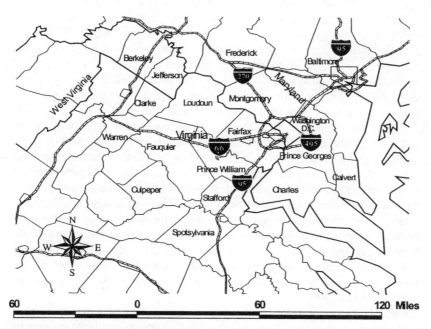

Figure 1.1   The Washington Metropolitan Region

throughout the remaining chapters in an effort to reduce redundancy and confusion in other parts of the book.

ORGANIZATION OF THE BOOK AND THE CASE STUDIES

As described above, the second chapter of the book identifies and assesses challenges to the evaluation of ITS deployment projects and recommends a multi-criterion approach in the form of a checklist (see Appendix, Chapter 2). The remaining chapters of the book use the multi-criterion template as a basis for evaluating seven specific applications or application contexts of ITS in the US National Capital region. Chapters 3 through 6 describe the analysis and results of an interpretive application of the methodology and together these form Part I of the book.

Chapter 3, by Roberto Mazzoleni, is an evaluation of the deployment of electronic tolling systems (ETS) in two in-line segments of interstate quality highway in the Dulles Corridor. An important aspect of this case is institutional challenges. One part of the ETS implementation is for the public part of the roadway (Dulles Tollroad) and the other is for the private roadway (Dulles Greenway). Managing institutional coordination problems between the public and private sectors to create a seamless tolling system across the two segments of the roadway is a central evaluation issue in this chapter.

In Chapter 4, Brien Benson examines driver responses to the many variable message signs (VMS) displayed and in operation in the Northern Virginia part of the National Capital Region. The ITS evaluation check list is used to surface VMS-related behavior and perceptions. The results of the analysis were used to guide the design of a survey that was then used to poll a random sample of area motorists. Results and lessons learned are identified and the ability to apply them in other settings is considered in this chapter.

The focus of the case study in Chapter 5 is the deployment of automatic (optical electronic computer assisted) truck rollover warning systems installed at three interstate exit ramps around the Capital Beltway in Virginia and Maryland. Michelle Sager evaluates the results of the implementation of these systems using the checklist in an interesting combination of *ex ante/ex post* evaluation. No rollover accidents due to excessive speed have occurred on the three exit ramps during more than five years since implementation. A cost-effectiveness analysis indicates that this deployment has been cost-effective. The Commonwealth of Virginia plans one new deployment of this system near the end of 2000.

Chapter 6 examines the development and outcomes of a nearly 20-year evolutionary deployment of an Advanced Traffic/Transportation Management System in Montgomery County (Maryland). In this chapter, Hans Klein uses the ITS evaluation checklist to emphasize the primary outcomes of this project. Significant among these is the conclusion that the evolutionary incremental implementation scheme that defines the project reduced institutional barriers and facilitated learning. This has made it possible to extend the ATMS from supporting traffic-specific services (such as signalization) to others such as police, fire and emergency management.

In Part II of the book, the case studies use the checklist as a guide to the overall evaluation, but in addition develop and apply alternative methodologies to specific problems to a much greater degree than those in Part I. Laurie Schintler in Chapter 7 evaluates the planned deployment of an Advanced Transit Management and Deployment System called Smart Flexible Integrated Real-time Enhanced System (SaFIRES). SaFIRES was deployed in Prince William County (Virginia) on the south side of the region (see Figure 1.1). The study focuses on the deployment of a system that is to provide transit-type services to a population that resides in a highly dispersed suburban community on the periphery of the region. Results of the analysis inform the deployment of other systems designed to provide transit services in low-density suburban environments. One interesting extension of the checklist tool used in this study is the use of a planning balance sheet methodology.

The Woodrow Wilson Bridge is a major link in the East Coast I-95 corridor. It is currently in poor repair and is to be replaced in the next 5–7 years with a higher-capacity bridge. Three technologies were identified as potentially useful for enhancing congestion management, safety management and mobility with respect to the new bridge: video monitoring; automated tolling; and information devices to help drivers manage congestion and delay. While using the checklist methodology to evaluate potential outcome effects of adding each of these technologies to the bridge design, the authors, Kingsley E. Haynes, William Bowen and Laurie Schintler, extend the analysis and use three alternative methodologies to investigate specific ITS implementation problems. The first methodology, probabilistic multidimensional scaling (PROSCAL) is used to assess goals and preferences of and for ITS technologies on the bridge. Meta analysis is used to investigate what is known about pricing and traffic demand (elasticity) in an effort to inform the utilization of electronic pricing possibilities on the bridge. Finally, bottleneck modeling is used to simulate the impact of pricing and congestion on the timing and potential route choice behavior of Woodrow Wilson Bridge users. In short, these methodologies are used to probe, *ex ante*, into questions about how the bridge will operate and be managed.

Chapter 9, the final chapter, by Gerard Maas, Mark Maggio, Hadi Shafie and Roger Stough, applies simulation modeling to estimate the effects the addition of several ITS technologies to current incident management measures in the Northern Virginia region will have on congestion and mobility in

Northern Virginia. A model based on national metropolitan traffic flow patterns was used to estimate what the congestion mitigation effects of the ITS components would be when added to the incident management service currently in operation. The significance of this case study is that it illustrates how a simulation methodology developed for one purpose can be reconfigured for another. It also demonstrates convincingly one way that ITS may increase the productivity of existing roadways and estimates time savings or delay reduction that may result.

NOTES

1. Data from the US Bureau of the Census, Census of the Population for 1970, 1980 and 1990 all evidence continued and accelerating dispersal of residential locations and of work places (see also Ewing, 1994; Linneman and Summers, 1991; Stanback, 1991). Further, following each of these census definitions most major metropolitan regions have expanded to include new peripheral jurisdictions. For example, the Washington, DC Standard Metropolitan Statistical Area increased from 15 jurisdictions in 1980 to 23 in 1990. The 1990 definition included jurisdictions in Maryland, Virginia, two counties in West Virginia and the District of Columbia. Additional counties are to be added in the 2000 definition.
2. For example, the Federal Communications Commission recently ruled to make the adoption of a nationwide 511 traffic information telephone number possible.
3. ITS can improve intermodality through improved communication systems such as ATIS and ATMS. Improved safety can be achieved with, for example, collision avoidance systems. Further, through ITS traffic flow can be smoothed out, thereby increasing capacity while at the same time reducing emissions and therefore improving air quality.
4. Although some are widely deployed, such as smart signalization at the intersection level if not so often at the system-wide level. Further, in-vehicle systems are increasingly becoming available on top-end vehicles, for example, GPS and security systems.
5. This knowledge gap is constraining diffusion and deployment of ITS at the local regional level.
6. Although, for example, Houston, Los Angeles and the Greater Washington Region are in various stages of deploying such systems.

REFERENCES

Bourne, L.S. (1992), "Self-fulfilling prophecies: Decentralization, inner city decline, and the quality of urban life", *Journal of the American Planning Association* **58**, 509–13.

Champion, A.G. (ed.) (1989), *Counterurbanization: The changing pace and nature of population deconcentration*, London: Edward Arnold.

Ewing, R. (1994), "Characteristics, causes, and effects of sprawl: A literature review", *Environmental and Urban Issues*, Winter 1–15.

Garreau, J. (1991), *Edge Cities: Life on the new frontier*, New York: Doubleday.

Gordon, P. and Richardson, H.W. (1996), "Beyond polycentricity: The dispersed metropolis, Los Angeles, 1970–1990", *Journal of the American Planning Association* **62**, 289–95.

Levernier, W. and Cushing, B. (1994), "A new look at the determinants of intra-metropolitan distribution of population and employment", *Urban Studies* **31**, 1391–406.

Linneman, P.D. and Summers, A.A. (1991), "Patterns and process of employment and population decentralization in the US, 1970–1987", Wharton Real Estate Center Working Paper 106.

Stanback, T. (1991), *The New Suburbanization: Challenge to the inner city*, Boulder, CO: Westview Press.

Stough, R.R. (1995), "Technology will spur satellite cities, more sprawl", *Edge City News* **3** (4).

Stough, R.R. and Rietveld, P. (1997), "Institutional issues in transportation systems", *Journal of Transportation Geography* **5**(3), 207–14.

Vance, J.E. (1990), *The Continuing City: Urban mythology in western civilization*, Baltimore, MD: Johns Hopkins University Press.

PART I

A Multi-criteria ITS Evaluation Methodology
and Applications

2. Methodological and Technical Challenges in Regional Evaluation of ITS: Induced and Direct Effects

Roger R. Stough, Mark E. Maggio and Dingjian Jin

INTRODUCTION

The last decade or so has witnessed the increasing application of computer and information technology to transportation infrastructure and vehicles. The resultant technical transport systems are called Intelligent Transportation Systems (ITS) in the US (Road Telematics in Europe). The continued development and application of these systems stems from a belief that ITS promises to increase the capacity and productivity of traditional transport infrastructure as well as contributing to the achievement of other goals such as safety. Consequently, the evaluation of its contribution to achieving transportation goals and objectives is of considerable importance and essential to determining if ITS is contributing to desired outcomes. There are a number of evaluation issues facing the successful assessment of the contribution of ITS. This chapter identifies and examines methodological and technical evaluation issues in the successful measurement of the effect of ITS on a diverse set of outcome dimensions. Further, it points toward the development of a new full cost (benefit) assessment methodology that is focused on the measurement of outcomes resulting from ITS investments.

This chapter identifies and examines methodological and technical evaluation issues in measuring the effect of Intelligent Transportation Systems (ITS) on a diverse set of outcome measures. The importance of ITS stems from a belief that it promises to increase the capacity and productivity of traditional transportation infrastructure. Consequently, the evaluation of its contribution to achieving transportation goals and objectives is of considerable importance.

The methodological and technical ITS evaluation issues are focused at the deployment level. Methodological challenges consider the viability of the assumptions adopted in conventional approaches to the evaluation of transportation investments. Technical challenges arise in the data, models and techniques that have not been able to overcome the shortcomings of conventional approaches to evaluation. In recognition of the complex and sophisticated processes at work in transportation and its interplay with other socio-economic sub-systems, these two types of issues are examined in the hope that a new evaluation framework may be adopted. While the issues have been separated into methodological and technical components in an effort to better manage the discussion, the difference between them is often blurred. For example, measurement is an issue in both types of challenges. An effort to minimize this redundancy has been made throughout this chapter.

This chapter is an attempt to broaden and deepen earlier work on the topic of ITS evaluation (Stough and Maggio, 1994). That work reviewed and assessed the conventional evaluation approaches to transportation systems effects of ITS investment, non-traditional and induced effects of ITS investment, and the methods available for a total-cost approach to ITS evaluation. The focus here lies with identifying the relevant and important outcome dimensions and with assessing and evaluating different measurement methodologies that are available for each of the outcome dimensions. As such, the chapter points toward the development of a framework that can guide the ITS evaluation process. The other chapters in this book apply the checklist evaluation framework provided at the end of this chapter.

ITS is not a single technology but a set of related information and communication technologies that are being applied to transportation infrastructure and vehicles. Most of the technologies already exist, that is, they are off-the-shelf technologies that have been reconfigured for application to the transport environment. In the literature on ITS, these technologies have been classified into application areas in an effort to simplify their description. These include: Advanced Transportation Information Systems (ATIS); Advanced Traffic Management Systems (ATMS); Commercial Vehicle Operations (CVO); Collision Avoidance Systems (CAO); Variable Message Signs (VMS); Electronic Toll Collection (ETC); and Electronic Guidance Systems (EGS) or "drive-by-wire", to name a few. The specific technologies include but are not limited to electronic sensors, wire and wireless communication, Global Positioning Systems (GPS), software, hardware, and systems design and integration.

Many ITS technologies such as ATIS and ATMS must be implemented at the functional regional (for example, metropolitan region) or corridor level to achieve maximum benefits. While this may be problematic from an intergovernmental relations perspective, traditional transport infrastructure for the most part is provided and managed at either the state or local jurisdiction level, not the cross-jurisdictional, regional or corridor level; it is the reality of ITS systems that they must usually be provided on a region-wide (corridor) basis for optimal results. For this reason, this chapter considers the metropolitan region as the primary unit of analysis. A case could be made for making the corridor the unit of analysis also, but this is not pursued here.

In any attempt to evaluate a program or policy, it is important to consider who the client is for the evaluation. This is an extremely important question for transportation in that there are many communities of interest and the list has been growing (Stough and Rietveld, 1997), and different interests have different sets of costs and benefits they consider important. For example, local taxing authorities will have a different view of what is important in the evaluation than drivers or auto manufacturers. Manufacturers of ITS equipment would have yet a different view. While this is a fundamental issue, it is not dealt with directly in this chapter other than to recognize that it is important to understand that there are different views and that any evaluation must itself be assessed from the perspective from which the analysis was conducted. What this chapter does do, however, is to internalize a number of evaluation components that have traditionally resided outside standard transportation assessments to ensure that interests which would have otherwise been left out of the analysis are included. Consequently, evaluation dimensions for the environment, social equity, economic development and institutional issues have been added to ensure that any evaluation using the proposed framework must at least consider these non-traditional dimensions and their associated communities of interest.

Context

Intelligent Transportation Systems exhibit special intersectoral and intertemporal effects that require the adoption of a non-traditional, total-cost (and benefits) evaluation approach. Because of the complex and intertwined nature of ITS with other transport and societal systems, the effects of its deployment are cross-sectoral and intertemporal in nature. Yet a review of research on the interplay of land use, transportation, energy, and emissions, among other factors, found no integrated urban model that provides an adequate analytical framework for capturing both the qualitative and quantitative measures of transport policy or investment outcomes (Anderson *et al.*, 1994, p. 6).

Traditional transportation benefit assessments have focused primarily on the immediate impact of policy within the transportation system (for example, improved freeway lighting leading to a reduction in freeway accidents). However, the transportation system is just one element in the highly complex, interdependent metropolitan interaction system. And in light of the high stakes (reduced congestion, improved competitiveness) and the irreversibility of some infrastructure decisions, any effort to evaluate the contribution of various ITS investments must accommodate a long-term vision of the metropolitan area. Consequently, an outcomes analysis approach, based on projected multi-modal and regional network effects, has been adopted here.

The built-up metropolitan transportation system has "locked-in" many individual and institutional behaviors and researchers suggest that people will continue to behave in the same manner as they do today. In addition to this behavioral lock-in, the physical infrastructure, the bulk of the transportation system, is also locked-in, as it exists today. At the same time, resources for transportation investments are scarce. Yet the transportation system is increasingly called upon to deal with a number of unexpected problems: increased international economic competition; energy constraints; environmental damage; time wasted in congestion; transportation-related deaths and injuries; urban sprawl; and exclusion or limited access for the nation's poor (see Schulz, 1991, p. 12; Stough and Rietveld, 1997).

Although changing individual behavior and building new capacity may be difficult, it is possible to add information and communications technology to improve the existing system. The commercialization and conversion of defense technology has accelerated the development of ITS technology. This development, driven largely by advances in information processing, communications, and vehicle technology, has met with and contributed to ever-increasing surface street and highway congestion (through its growth-inducing effects), concern for safety, and maintenance.

ITS will be most heavily deployed in metropolitan areas, that is, on a regional scale, where important safety and congestion benefits could be best realized.[1] Given that the majority of people live in or adjacent to metropolitan areas, peak (rush) hour demand for ITS as well as other improvements is high. With an increasingly greater proportion of metropolitan traffic crossing municipal, county, or state borders, the need for multi-jurisdictional applications of ITS in metropolitan areas is pressing.

Outcomes Analysis and the Total-cost Approach

In approaching the ITS assessment problem, an "outcomes analysis" framework has been adopted. This means that the focus of interest is on the measurement of outcomes (positive, negative, direct, indirect and induced) resulting from ITS and related investments. This approach is consistent with the total-cost approach which has been proposed for evaluating transportation investments and operations (Green *et al.*, 1997; Murphy and Delucchi, 1998; Mudge, 1994; MacKenzie, 1994; Gifford *et al.*, 1994; DeLuchi, 1993; Mudge, 1992). A total-cost evaluation permits the full inclusion of the external effects of a transportation investment. Publicly optimal economic policy based on the consideration of these project or investment "externalities" has been clearly established (Arrow and Kurz, 1970).

A total-cost approach takes into account the environmental, behavioral, and jurisdictional benefits or costs associated with an ITS investment (Horan, 1992; Horan and Gifford, 1993). This approach addresses the complex set of human interactions "which are the ultimate shapers" of transportation, urban form and the environment (Pisarski, 1991), while conventional methods tend to be too mechanistic in their assumptions about the linkages between transportation and the environment and metropolitan form.

In transit systems evaluation, standard measures (for example, new riders, cost per rider, and fare revenue generated) are being augmented by total-cost concepts, including air pollution reduction, automobile congestion reduction, and other non-economic benefits. A modified total-cost approach has been applied to a large European infrastructure project (Lorenzini, 1994), and significant elements of the total-cost approach have also been suggested for inclusion in a model for all of Europe (*Investment ...* , 1992).

The importance of the total-cost approach in accurately reflecting modal balances is illustrated by the high level of automobile subsidies. A recommendation by Pucher and Hirschman (1993) suggested that automobile users pay the full social, economic, and environmental costs of driving. These costs include roadway congestion, air pollution, noise, traffic accidents, energy consumption, parking problems, as well as road construction and maintenance costs.[2]

The fate of transit properties, and particularly movement toward a balanced, total-cost approach to modal subsidies, is largely outside the control of transit authorities and advocates. The massive political pressures to keep automobile use inexpensive will continue to dominate the debate (Pucher and Hirschman, 1993).

In this chapter, the focus is primarily on the indirect, less traditional effects (outcomes) of transportation investments, many of which are of current policy interest, such as the effects of economic development. However, the direct or more traditional in-system effects of transportation investments, such as safety outcomes, congestion outcomes, construction cost, and maintenance cost are still important.

In a transportation context, and in some cases in a broader policy context, several linked outcome effects – such as employment, income, output, private-sector productivity, fiscal considerations, environmental quality (as well as other quality of life considerations), social equity, and effects on institutions – are vital. Taken together, these traditional and non-traditional outcome dimensions provide the foundation for a methodology for metropolitan region evaluation of induced and direct effects of ITS and other transportation investments. However, there are a number of significant methodological and technical challenges in the evaluation of these ITS investments.

Methodological Challenges

To examine, in depth, some of the weaknesses inherent in more conventional approaches to evaluation, and to move toward a total-cost approach, a set of methodological challenges are identified in Table 2.1. These are the ten principal methodological challenges for the estimation of direct and indirect effects of ITS as well as transportation investments in general.

Time horizon
In estimating the effects of ITS and transportation policy initiatives, especially induced effects, often the appropriate time horizon is quite distant. The extent of this period creates difficulties in the use of marginal analysis (which by nature is incremental) because many intervening conditions will subsequently change. When estimating effects, the time horizons that are employed must be carefully chosen according to the nature of each effect.

Table 2.1   Methodological challenges

Time horizon
Vision
Increasing returns
Demand estimation and travel demand models
Induced demand
Broad context of induced effects
Functional regions
Subjectivity
ITS as an innovation
Exactness

Induced effects usually take longer to appear, as the impact of transportation infrastructure has always had a long time horizon. However, when longer time horizons are considered, marginal analysis is no longer effective, as it does not accurately account for intervening factors. As a result, conventional cost-benefit analysis based on marginal analysis is inadequate (Stough and Haynes, 1997).

In an on-going investigation of the contributions of transportation investment to economic productivity, it is found that most research neglects long-term transportation impacts on development, which involve large-scale socio-economic transformations (Jones, 1995, p. 1).

A longer time horizon is also indicated by findings which suggest "the lag in the economy's adjustment to new transportation infrastructure [is] from five to ten years or even longer" (ibid., p. 29).

Vision

For a time horizon that is longer than 10 years, typical for transportation infrastructure, it is nearly impossible to evaluate transportation or ITS projects without a vision of the future of the region. It cannot be assumed that induced changes are always predetermined by the past. Nor can it be assumed that a region can arbitrarily bring about whatever changes it desires. But a region may have some influence on the direction of change if it investigates the nature of changes, and if it has a clear vision of its desired future. The vision of the future is the goal that guides decision making and outcome achievement.

A variety of metropolitan strategic planning approaches have been developed over the past 10 years that can be used to facilitate vision development and implementation. Methodologies for facilitating the visioning process include churettes, Nominal Group Technique (Delp, 1977), and scenario planning. More recently Roberts and Stimson (1998) and Stough *et al.* (2000) have developed a method for visioning and implementation planning called multi-sector analysis (MSA). The MSA approach provides a method for defining and managing groups in the visioning process while at the same time aiding them in creating development scenarios and implementation plans. Further, the recent interest in industrial cluster analysis as a component to help inform scenario analysis and visioning is important (Feser and Bergman, 2000; Hill and Brennan, 2000; Porter, 2000; Stough *et al.,* 2000).

Vision along with development scenarios are indispensable in evaluating the consequences of transportation projects. The lack of a clearly defined long-term vision of regional conditions (and a strategy to get there) reduces the predictive power of any model. In the absence of vision, the analyst is left with few clues as to the direction and magnitude of pertinent structural changes that will affect transportation.

Increasing returns

Conventional cost-benefit analyses in transportation planning and evaluation reflect a Newtonian world view, which until recently dominated economic thought. With this view, marginal changes in the input to a system can only cause marginal or incremental output changes.

However, according to the science of complexity theory, most socio-economic systems are adaptive, self-organizing, and autocatalytic. Within these systems, positive feedback produces the phenomenon of "increasing returns" (Arthur, 1990; Waldrop, 1992). In complex systems, a marginal change in the input may cause great changes (or no change) in output (Rycroft and Kash, 1999).

Elements within these systems form autocatalytic sets that cause perpetual novelty and evolution. Within these sets, the existence of any one element becomes indispensable to the effective operation of the other elements. Consequently, the marginal effect of a single element on the system is considered both impossible and useless. Analysis must focus on the elements within the autocatalytic sets as a whole. In short, the autocatalytic set becomes an important policy and outcome related variable.

Operationalizing the concept of autocatalytic sets is critical to improved research results. The configuration of the autocatalytic set within a specific metropolitan area must be identified. The system must track the elements in the group or "set" of variables to determine the nature and the type of feedbacks, and how they interact and operate.

From this world view, one may visualize the transportation system as one of the elements that constitutes the real or potential autocatalytic set for regional development or for environmental sustainability. Ultimately, the role of the transportation system can only be understood by analyzing the nature of the autocatalytic set.

To address the challenge presented by the presence of increasing returns, an understanding of different types of autocatalytic sets in regional development and the different elements within these sets needs to be developed. The elements of an autocatalytic set which may promote regional economic development include:

- technological infrastructure;
- transportation and telecommunication infrastructure;
- business networks;
- informational infrastructure;
- institutional infrastructure;
- industrial economic structure;
- environmental and energy related components.

Demand estimation

Most transportation-related models rely on future estimates of transportation demand. However, estimation of demand has proven to be perplexing. Demand is difficult to estimate because traveler reaction and behavior to various policy changes is not well understood. Also, demand estimates using elasticities are unidirectional; they do not take account of numerous two-way interactions and related feedbacks.

Furthermore, there is a gap in our understanding of some very fundamental relationships between transportation and other socio-demographic factors. This is manifested in "weak perceptions" of the preferences and the relationship between household activity and travel demand (Pisarski, 1991, p. 6). This gap in our understanding persists.

The research needs for improved estimates of travel (identified by the Federal Highway Administration in Brand 1991, p. 114) are summarized below.

Despite the fact that this list is nearly 10 years old, and that some additional knowledge and understanding have been acquired, these research needs remain largely unmet. They are:

- how individuals react to information, travel conditions, and costs in their decisions to consume land and travel;
- how resources and travel conditions are likely to change in the future with respect to individual behavior (for example, constrained parking ability, rising housing costs or remote work such as telework options) and how these changes are likely to affect congestion;
- how populations and individual needs are likely to change in the future;
- how new technologies will affect demand in different ways than current technology; and
- what transportation options can both improve travel conditions and reduce travel demand, if any.

There are a number of widely used travel demand estimation models. For example, TRANPLAN enables a metropolitan planning organization to accomplish long-term transportation planning. Each link of a regional transport network is coded according to its characteristics and capacity, and loaded onto a simulated network with respect to current conditions and with respect to some alternative future conditions. The outputs of TRANPLAN include: delay times, vehicle-miles, and average speeds by network segment.

The Highway Land-use Forecasting Model II+ (which runs in a Windows environment) estimates the relocation of population and service employment brought about by changes in a highway or transit system. Population and employment reallocations are handled by the widely accepted Lowry–Garin model of land use. Models of this type could be incorporated into regional input–output econometric models. A resource that lists and describes hundreds of transportation models has been produced by the University of Florida (1999–2000).

Induced demand
Another methodological challenge related to demand estimation is the chronic problem of induced demand (Button, 1982). The practice of transportation modeling has long faced the problem of excess latent demand, or induced demand. Experience has repeatedly demonstrated that capacity expansion is usually accompanied by increased demand or a modal or route shift. Although demand can sometimes be satisfied on particular road segments or at bottleneck areas, the magnitude and resiliency of latent demand that appears across systems is considerable. This expression of latent demand was first described more than 30 years ago (Downs, 1962) and more recently (Downs, 1992).

The vision of ITS is that it can create additional transport capacity. To the extent it does this through network efficiency gains, it will have an effect similar to that of an increase in traditional capacity, for example, a new lane of roadway. Consequently, ITS will most certainly contribute to induced demand and, therefore, evaluation approaches must estimate this effect. However, conventional methods and models for estimating induced demand from lane or roadway expansion lack precision. Applying these methods and models to estimate the induced demand of ITS improvements is in a nascent stage of development. However, some work is underway and appears in the case study Chapters 8 and 9 of this book.

The broad context of induced effect

Analysts face a predicament in developing appropriate algorithms to estimate second- and third-level indirect impacts, as there is not a full understanding of the functional and empirical relationships between the effects of transportation policy and the various social and economic reactions to changed circumstances.

There may be additional intervening (omitted) variables which influence the relationships under study. As all conceivable variables cannot be included in a model of this type, it is important to construct the model with this in mind by including those variables which are essential.

Often referred to as the "indivisibility" issue, researchers may have difficulty in deciding which variables to include. In examining incremental changes, marginal analysis may be used to decide which variables are more important. However, as these estimated relationships do not remain static in the face of numerous multiple changes, researchers are often unsure which factors are the most important.

Additionally, a more fundamental issue is the ability of operational models to take into consideration the recursive effects of related transportation phenomena. Their ability to assess the extent to which such recursive feedback occurs has been questioned, and many models may lag behind the state-of-the-art (Harvey and Deakin, 1993, pp. 3–70; Deakin, 1991, p. 38).

Functional economic regions

In transportation modeling approaches, data are often not collected or analyzed on the basis of functional economic regions. Such regions may be defined as spatial areas with integrated settlement patterns, measured by population density, growth, or urban population characteristics. A functional region acknowledges the importance of the geography of journey-to-work patterns (for example, areas are joined by a road network of a given magnitude and connectivity) and can be defined in terms of a specified level of commuting among a set of political jurisdictions that define a labor market.

Functional economic units may also share a predominant production mode, and are necessarily geographically contiguous.

Problems with indirect effect estimation become more pronounced as many reactions and recursive effects are region-specific, not national. There are problems not only in obtaining regional data, but in determining which causal factors have precipitated these reactions. One must also anticipate the effects of a neighboring region's policies.

Additionally, while the total-cost approach can be applied to a number of metropolitan regions, a specific model will not be transferable to another metropolitan region without recalibration and alteration of its underlying set of structural econometric equations. This stems from the heterogeneity of regions. Each transportation system is different, from the sectoral make-up of its production systems to its structural linkages to transport.

Functional economic regions are important for the evaluation of ITS because they define geographic communities of interest. Thus, such regions define the ITS implementation context that is likely to generate maximum benefits because of the synergistic effects efficient transport has upon all functions in the region. In short, the functional region is the deployment unit that will generate the most benefits. This implies that diffusion of ITS is for the most part going to take place at the functional region level. To date the deployment of ITS has tended to support this conclusion. While federal and state governments have played a role (for example, in standards setting and national systems architecture), the implementation of ITS systems have for the most part occurred at the functional regional level.

Subjectivity

To aggregate all measures of a project's direct and indirect effects into one indicator of total project desirability, analysts have either monetized all the effects into dollar values, or they have used standard units of utility (utile) to measure combined effects. However, these methods inject a significant degree of subjectivity into the evaluation process. Even with multi-criteria analysis, where projects are assigned rankings, effects must have weights assigned – another subjective process.

Subjectivity is an issue in every aspect of the evaluation process, not just the unit of measurement used or the weights applied. There is always an issue of who or what communities of interest have standing in the evaluation process and more specifically who the evaluation is for. Further, the selection of methods, measurement techniques and levels of coverage in the analysis are themselves ultimately subjective in nature. In short, subjectivity is ultimately the central issue in any evaluation. It is incumbent on the researcher and those who consume or use an evaluation to be overtly vigilant in articulating where subjectivity creeps into the evaluation. This is particularly important in the evaluation of ITS because it is seen as a bundle of new technologies offering heretofore unrealized benefits. The articulation and recognition of subjectivity could easily be lost in the perception of the promise of this new technology.

ITS as an innovation

Since many ITS systems use innovative technological applications that have not yet been broadly implemented, good sample data on the effects of these improvements do not exist. Conventional transportation models may be incompatible as they are often predicated on issues of capacity change rather than changes in information flow and communications improvement, as we see with ITS (Harvey and Deakin, 1993, pp. 3–6). However, there are uncertainties in the effects of new ITS processes and procedures that augment efficiency, as opposed to simple capacity expansion.

Exactness

Modeling practice has often adopted sweeping assumptions to circumvent many of the aforementioned challenges. This results in a loss of precision, and in many cases these models cannot provide exact estimates. Despite their diminished accuracy, such models may provide a heuristic tool and a decision making framework for the multivariable comparison of alternatives.

These methodological challenges are common to the estimation of both direct and induced effects. Moreover, as noted in the introduction, these challenges will overlap with the technical challenges and, therefore, will be found throughout the analysis of all of the nine technical outcome dimensions.

TECHNICAL CHALLENGES

Conventional Methods of Evaluating ITS Investments

To date, there has been a tendency to measure the anticipated effects of public investments quite narrowly, in terms of fairly obvious and conventional outcomes, using the well-developed cost-benefit framework. In this approach, the time-profile of costs and benefits is established for every alternative project and transportation investment. This cost-benefit method is the traditional method of determining the worth, value or feasibility of a public investment. Discounted present value (DPV) reduces costs and benefits of an investment stream or benefit stream to a single figure at a common point in time. By comparing the DPV of competing projects, those which meet the Pareto-optimal efficiency criterion (assuming Kaldor–Hicks compensation, see Hicks, 1946) can be pursued.

Additions or reductions in social welfare, which can be expressed in dollar terms, should be included in a cost-benefit analysis (Hotchkiss, 1977). Benefits include various concepts of income, receipts, time value, subsidies, and economic utility. Benefits accrue to users and non-users alike. Negative externalities (disamenities) must be subtracted as costs. Spill-over effects of a project must be evaluated and included. Spill-over effects occur when the actions of one economic agent affect another agent directly, but not through price transactions (for example, when a business location decision causes increased traffic congestion). The result, after monetizing all the effects (including spill-over effects), of social benefits minus social costs equals the net social benefit.

In practice, public investment project estimates often include net present values or internal rates of return. In theory, discounted present value should include consideration of opportunity cost, but this is rare in practice. Serious analyses or discussion of these considerations (for example, opportunity costs, social equity, or institutional adjustment costs) are rare, indicating that these indirect effects are not an integral part of conventional transportation infrastructure investment analyses.

With these induced effects, transportation infrastructure is of course necessary for regional development, but it is not the only condition required for regional development. There are other factors crucial in influencing the nature and magnitude of any induced effect. One method that identifies the contribution of a transportation project to an induced effect is marginal analysis. Marginal analysis operates under the assumption that other factors are held constant – usually this is done by estimating price or cost elasticity.

However, this method is subject to two constraints. Primarily, marginal analysis relies on the conditions of factor availability at a specific location. Due to wide differences in factor endowment, there may be great divergence in the results of the estimation. Widely observed divergence in the social rate of return to transportation infrastructure reflects this difference. Therefore, data and estimates from other regions cannot be used to justify local transportation projects. To be effective, data used in estimation must be region-specific, as noted in the "Methodological Challenges" section above.

The second constraint is that while the marginal effect of a transportation project may provide adequate justification for a particular project, investment in other (non-transportation) factors may actually be more effective and efficient in inducing the desired effect (for example, economic development and social equity). In addition to considering the marginal effects of a transportation project, estimates of the marginal effect of other projects aimed at economic improvement are needed, that is, "opportunity cost".

There are several areas of direct and induced effects, or outcome dimensions, which must be considered in a total-cost approach to the evaluation of

ITS or transportation investments (see Table 2.2). These outcome dimensions include ones based on traditional transportation measures (safety, congestion, construction costs) or on induced effects (economic outcomes, environmental effects, fiscal effects), and those with qualitative measures (equity and institutional effects). A discussion of the technical measurement problems that may be encountered in each of the nine outcome dimensions, along with a brief review of available or recommended models for each, is provided below.

Table 2.2   Technological challenges

Safety outcomes
Congestion and delay-related outcomes
Capacity, mobility and network outcomes
Construction and installation cost outcomes
Economic outcomes (employment, income and output)
Fiscal outcomes
Environmental outcomes
Equity outcomes (quantitative and qualitative)
Institutional outcomes

Safety outcomes
Safety has been a traditional concern in transportation investment analysis. The goal has been to choose investments in infrastructure and operations that reduce traffic related deaths and injury, and property (vehicle) damage. Factors ranging from road geometrics in the design of roadways to signalization have been considered as part of cost estimation. In the future, it will be important to consider the inclusion of ITS technologies which may affect safety, such as collision-avoidance systems, automated highway systems, ATMS, and advanced vehicle inspections.

Safety databases, although they cannot estimate or predict the effects of alternative changes to the highway system, are widely used today. They include packages such as ACCISUM, HISAM, and SCARS that are distinct from the available accident modeling packages.

There are a number of widely available MPO-level accident (safety) models (for example, HISAFE) that evaluate the effectiveness of accident countermeasures following implementation. This model determines the expected change in the accident rate, based on proposed change(s) to a high accident location. HISAFE can also be used to perform economic analyses of several competing alternatives (University of Florida, 1999–2000).

Congestion-related outcomes

Congestion has also been a traditional concern. The nation's transportation systems are in many cases operating at or in excess of their design capacity. Many observers find the road network in the United States (approaching 4 million miles) to be in poor physical condition and congested, relative to earlier periods (such as the 1970s). Roughly half of the non-local road miles are classified as "fair" or "poor" by the Federal Highway Administration.[3]

In 1987, 46 percent of urban highway miles were classified by the Federal Highway Administration as "congested" – 65 percent during peak periods. Although there is a body of literature (Hartgen and Krauss, 1993; Tatom, 1993) which argues that the system is not in poor condition, the evidence of congestion in metropolitan areas is evident. Vehicle miles of travel are growing in some regions at a rate of 5 percent annually without compensating increases in lane mileage. Metropolitan freeway delays from congestion are expected to reach 4 billion hours annually by 2005 – urban highway delay is forecast to increase 436 percent over the period 1985–2005 (US Department of Transportation, 1987). Clearly, congestion reduction is an important outcome measure for any ITS evaluation.

Concern with "gridlock" and congestion has been a primary outcome consideration for some time. Measurement has focused on such attributes as travel times, personal mobility, accessibility, total vehicle-miles of travel (VMT) as well as other related measures. One ITS technology, electronic toll collection (ETC), makes road pricing (particularly, variable pricing) a feasible way of allocating scarce road infrastructure, thereby affecting congestion levels. Achieving desired congestion outcomes in a metropolitan context in the future will require the serious consideration of this ITS possibility as well as other applications such as Advanced Traveler Information Systems (ATIS) and Advanced Traffic Management Systems (ATMS).

Congestion-related outcomes can be measured in terms of travel times, average hours of delay, or average speeds on a given corridor or on a highway system in a particular area or region. Efforts to measure the costs of delay and congestion have become increasingly important. Researchers and planners are working on congestion pricing in order to estimate the social cost of congestion. These estimates may eventually yield an "optimal" congestion price (or user charge) for peak period highway use (Mohring and Anderson, 1994). If accurate estimates of marginal cost user charges can be developed, they can be used to compare the value of alternative projects. These cost estimates may be derived by trial and error, through willingness-to-pay studies, or through examination of revealed preferences. Ultimately, ITS applications could reduce the cost of implementing a congestion pricing system.

A widely used and well-accepted congestion model, FRE-Q,[4] uses capacity additions and constraints to measure the effects of delay on a transportation system. FRE-Q produces estimates of delay, average speeds, variation in speeds, and correlates the data with accident levels. A similar model could eventually be used as a sub-component in a larger regional input–output econometric model to provide congestion estimates to be used in the sectoral input–output matrix.

Capacity, mobility, and network outcomes

Capacity and mobility outcomes are closely linked to other traditional transportation system objectives, namely safety improvements and congestion mitigation. Capacity and mobility may be measured with respect to vehicle-miles of travel (VMT), ease of personal mobility, or system accessibility. Typically, these outcomes may be measured and evaluated quantitatively. However, qualitative measurement and evaluation of any proposed ITS investment for its adaptability and flexibility for future system capacity changes must be considered.

In an era where rapidly growing transportation demands are constrained by limited public budgets, it is important to examine the intermodal and network efficiency potential of ITS investments. Future projects must work to produce seamless networks and smooth linkages at transfer points and terminals. Networks need to be optimized. Improved public information (through ITS) on modes, routes, and system status will lead to more efficient capacity utilization.

Traditionally, the effects of any transportation system improvement on the overall highway network – or on passenger or freight intermodal linkages – have been evaluated using mathematical network or flow models; although methods to optimize network interfaces, at terminal and transfer points, between modes, and in intermodal systems with significant "noise" or system uncertainty, have been scarce. Current transportation investment analyses provide some evaluation of the intermodal impacts, through modal shift analyses and cross-elastic price estimation. However, the intermodal assessment is sometimes relegated to an afterthought, rarely more than a paragraph at the end of the assessment report.

Federal legislation, such as the Intermodal Surface Transportation Efficiency Act (ISTEA) and its successor, the Transportation Equity Act of 2000 (TEA21), stresses the importance of improving intermodal linkages through multi-modal planning of transportation investments. A combination of analytical processes and value judgments will be necessary to achieve the proper mix (highway/rail/transit) under ISTEA/TEA21 and planning flexibility. Under the system of open competition, metropolitan areas may have to place a high value on a project's potential social, energy, and environmental effects.

Federal funds for state transport departments are no longer lump sums determined by formula according to each state's share. Since TEA21, as its predecessor, ISTEA, has precluded a purely "common pot" approach to modal or sectoral planning, a balance between freight/passenger, urban/rural, motorized/pedestrian, and highway/transit projects is now required – and has been sought (Covil *et al.*, 1994). Any integrated approach to evaluating alternative transportation investments, including ITS, must attempt to measure the contribution of the investment to intermodal linkages, and to optimizing the development and spread of seamless transportation networks.

Evaluation of the intermodal and network effects of a given ITS or other transportation investment will, by necessity, include quantitative as well as qualitative factors. Because intermodal linkages and network adequacy are not subject to commonly accepted or standard units of measurement, some variant of multi-criteria analysis will need to be adopted to evaluate which projects are more promising.

Construction and installation cost outcomes
The traditional approach to assessing the effects of transportation infrastructure, maintenance and operations employs conventional measurements. The goal for this part of the model assessment methodology will be to extend the application of the measurement dimensions shown in Table 2.2 to ITS components. This will require the application of traditional cost assessment and estimation procedures to potential or proposed ITS investment components, whether singular or in combination with other transportation improvements.

User benefits associated with highway improvements can be estimated and modeled by a number of available software packages such as Highway User Cost Accounting, which adopts techniques from the 1985 *Highway Capacity Manual*, the 1977 AASHTO *Manual of User Benefit Analysis*, the 1982 ITE *Traffic and Transportation Engineering Handbook*, and the 1992 ITE *Traffic Engineering Handbook*. This package is useful in alternatives testing, and it is not a design tool (University of Florida, 1999–2000).

Economic outcomes
Many ITS investments will produce linked economic effects in terms of employment, income, and output. However, these are often long-term effects, and the acquisition of data and the measurement of effects in a metropolitan area may be difficult. Furthermore, ITS investments may only partially contribute to increased employment, income, or output in the metro area. These factors contribute to the difficulties in measuring economic outcomes of any ITS investment. Typically in such cases, an input–output econometric model or a production function approach is adopted.

There have been a number of studies that have examined the contributions of public infrastructure (capital stocks) to economic output using production function analysis on a national level (Holtz-Eakin, 1988; Aschauer, 1989;

Munnell, 1990a) and disaggregated by states (Costa *et al.*, 1987; Eisner, 1991; Munnell, 1990b). The output elasticity of public capital has also been estimated for Japanese regions (Mera, 1973), for France (Prud'homme, 1993), and in a European context (Biehl, 1991), and for US regions (Duffy-Deno and Eberts, 1989; Eberts, 1986 and 1991).

However, such analyses have not been a part of transportation investment (project) analyses. When transportation policy-makers compare alternative possible investments, they do not now have information on the differing contributions to output or productivity expected from these potential alternatives.

When estimating the economic outcomes of a transportation project, production function analysis on a national or state level is not appropriate because, as noted in the introductory section, the impacts of all transportation projects are primarily region-specific and dependent on particular regional and interregional socio-economic conditions. Using the production function approach, different economists have obtained widely varying results of the effect of public sector capital on productivity. This is termed the "productivity puzzle" (Holtz-Eakin, 1992).[5]

Metropolitan regional input–output analysis is more appropriate than the production function approach in estimating these induced effects for several reasons ranging from better estimation capability to the ability to measure effects across multiple sectors of the regional economy. Further, linkage of the input–output model to an econometric component would permit the forecasting of some economic effects of the transportation project thus enabling

Table 2.3   Indicators of economic outcomes

Changes in output
  • gross regional products (GRP)
  • number of employment
  • income (total, and per capita)

Changes in input–output relations
  • input–output technical coefficients
  • productivity: time and cost saving
  • competitiveness: cost advantage, just-in-time delivery, positive externalities.

Changes in structure
  • sectoral structure
  • coefficients of export
  • coefficients of import
  • land use and spatial distribution

---

present value estimation.

When using an input–output model to estimate the economic outcomes of transportation projects, it is necessary to identify different indicators which reflect the economic impacts of the transportation investment. In Table 2.3, three major levels of changes in the input–output relations, induced by a transportation project, are identified. They are changes in output, changes in input–output relations, and changes in economic structure.

There are also three levels (orders) of causal effects through which an ITS transportation project affects economic outcomes. The first order effect of an ITS project is through the generation of investment demands that cause increases in gross regional product (GRP), employment and income. For example, the Northern Virginia Regional Econometric Input–output Model (NVREIM) can estimate these effects (Campbell and Stough, 1994) or any other such model (for example, see Hewings and Jensen, 1986; Leontief, 1966). Estimating the first-order effects is a fairly straightforward estimation procedure. The data requirements include:

- sectoral investment demands in project execution;
- the proportion of these demands that are realized in the region.

With second-order effects, ITS projects may reduce transportation costs in various sectors. Change of the sectoral technical coefficients arise from shifts in transportation sector inputs to other sectors, and from changes in other sector inputs to transportation. These changes of technical coefficients in turn produce shifts in:

- regional output;
- regional comparative advantage;
- regional economic and spatial structure.

Generally, other economic conditions are assumed to be fixed or "given," especially in the short run – 5 years. Estimating changes in regional output due to the changes of the technical industry-specific coefficients is relatively simple: use of an input–output model is sufficient. The data requirements include estimates of the level of improvement of sectoral productivity, which results from increased efficiency in the transportation system. This information permits estimation of the change in the technical coefficients in the input–output model. However, changes in regional comparative advantage resulting from the change in technical coefficients are relatively difficult to estimate. To do so, it is important to establish indicators and measures of regional competitiveness and develop region-specific inter-regional comparison data.

In the short run, the impact of the changes of the technical coefficients on the change in sectoral structure can be estimated by using the input–output table. However, in the long run, changes of the technical coefficients may cause third-order effects through a change in regional competitiveness. In this case, estimation will be very difficult, as the change in technical coefficients is only one of the factors that will cause regional structural change.[6] There will most likely be a number of other intervening factors that will cause structural changes, especially in the long run.

The third-order effects of an ITS transportation project usually occur through the following causal chain. Transportation improvements change regional competitiveness and attractiveness, which in turn change the location preferences of business firms. Immigration and emigration tendencies will also change, and this causes change in regional output, competitiveness, and the economic and spatial structure. This effect is most difficult to estimate, since it requires not only data on location preferences of business firms, but also the immigration and emigration tendencies of residents, and concrete and detailed information on local conditions that will influence these preferences and tendencies. There are migration models which could be adapted for this purpose, but because there are so many conditions and factors that influence these results, the effects of transportation improvements cannot be separated from the effects of other factors.

What is more, without a strong vision of the future, and with the phenomena of "increasing returns" and "path-dependence," a small change in the initial conditions could result in a dramatically different regional development pattern. In these cases, not only is marginal analysis inadequate, but regression techniques are also insufficient.

One way to avoid these traps is to use the concept of an "autocatalytic set" as a starting point to evaluate transportation and ITS projects. Rather than wrestling with the details of cost-benefit analysis, this method of analysis requires:

- a vision of the desirable future of the region;
- the infrastructure needed for this vision (including transportation, communication, industrial, technological, R&D, human resource, and governance);
- the sequence of activities and projects which can trigger movement toward the realization of the goals or vision.

In this case, rather than using the conventional modeling practice of economic estimation, which is based on an assumption of unchanging conditions of regional development, a broad vision of regional development must be created and incorporated. The analysis must be subject to a broad vision of possible change in the nature of regional dynamics and regional economic, environmental, demographic, and infrastructural conditions. Future work must bring the vision and the model together.

Fiscal outcomes

Some of the government revenues and expenditures related to an ITS investment can be directly measured through the current-account effects on transportation or public works agency appropriations and budgets, or on state and local government budgets generally. Those direct system effects can be measured with traditional accounting and cost-benefit methodologies.

However, many fiscal effects are often intergovernmental effects, especially in circumstances where federal funds are meant to be stimulative, rather than substituting for state or local spending, as with Federal Aid Highway Funds.[7] Due to the regional effects of national policies, fiscal effects may be disproportionately shifted from the nation to a particular region. Since states and localities have widely varying transportation revenue structures, ITS investments may not have the same fiscal effect in every state, region or jurisdiction. Through the process of unfunded mandates, the federal government has in cases shifted costly responsibilities to states and localities. These induced effects are not easily measured directly; however, a fiscal sub-model to a basic regional input–output model could greatly facilitate the estimation of the fiscal effects. Fiscal indicators are:

- local government revenue and expenditure changes;
- state government revenue and expenditure changes;
- federal government revenue and expenditure changes.

Environmental outcomes

Environmental consequence is one of the major indirect outcomes that has received considerable attention. No new highway investment, or even a major improvement, can be made without an environmental impact statement (EIS) or an environmental assessment (EA) and subsequent hearings to ensure public participation. At the same time, while environmental effects have become a part of the cost assessment process, they are not well-integrated into the traditional cost estimation process, nor are they linked to indirect economic effects. A residuals sub-model could be added to the input–output economic estimation procedure suggested above to accomplish a partial integration of the transportation cost estimation problem (see, for example, Lakshmanan and Nijkamp, 1980).

Transport planning for equity and environmental sustainability has been embraced in many European efforts. In that context, more comprehensive and intangible goals for transport planning, such as social and environmental goals, have been adopted. Further, tensions between the goals of efficiency (traditional) and equity (distributional or social) have become more obvious in the US. A third goal, sustainability, can be seen as the preservation of a long-term ecological–economic balance, or as inter-generational equity (Stough and Haynes, 1997; Masser *et al.*, 1993). Many environmental conditions (states) will have to be modified to move toward greater sustainability.

With the metropolitan focus of this analysis, primary concern with respect to environmental consequences rests with air quality considerations and particularly with non-attainment areas (although land use considerations and hazardous spills are also important to a total-cost approach) because many of the metropolitan areas have non-attainment status under rules for ambient air and ozone pursuant to the Clean Air Act Amendments of 1990 (CAAA).[8] This means that scheduled emission-reduction plans and their execution are required if a metropolitan area is to continue to receive federally appropriated highway funds under the ISTEA[9] and TEA21.

Such actions are perceived, in the short run, to have potentially growth-constraining impacts. However, it is envisioned that deployment of ITS technologies can assist metropolitan areas in meeting their environmental goals (Horan, 1993). For example, integrated traffic management and traffic information systems may smooth out traffic flow, thereby reducing congestion, fuel consumption, and regulated tailpipe emissions. Therefore, ITS can contribute to improved air quality and to improved metropolitan area competitiveness.

One way to assess the total contribution of an ITS investment to both economic development (competitiveness) and air quality is through a regional input–output model. This model must have an inter-industry emissions sub-model linked to the standard inter-industry economic transactions matrix. The SEAS model is an example of this type of linked model (Lakshmanan and Ratick, 1980). A more general discussion of the environment–energy–economy linked input–output models appears in Lakshmanan and Nijkamp (1980).

Environmental indicators
- Emission reduction: meeting air quality goals of Clean Air Act Amendments of 1990;
- number of wetlands or wildlife habitats crossed;
- soil erosion and run-off.

Factors in ITS projects that influence environmental outcomes by first-, second- and third-order effects may be schematically modeled as follows.

- *First-order effect:* ITS project => reduction of congestion => reduction of emissions given unchanged transportation demand.
  Evaluation method: decreased travel time multiplied by emissions per hour.
  Data required: change in travel time and emissions.

- *Second-order effect:* ITS project => reduction of congestion => increase in transport demand among the existing population => increase in emissions.
  Evaluation method: demand elasticity of transportation cost.
  Data required: relative change of transportation demand and cost.

- *Third-order effect I:* ITS project => reduction of congestion => increase attractiveness of the location => immigration of business and population => increase in transport demand => increase in emissions.
  Evaluation method: location or gravity model derived from a population or migration model.
  Data required: indicators of attractiveness and elasticity of migration.

- *Third-order effect II:* ITS project => increase in economic outcomes => increase in other sources of pollution, and increase in transportation demand => increase in emissions.
  Evaluation method: input–output model plus environmental model.
  Data requirement: emissions by industry matrix and industry inputs–outputs matrix.

Immediately upon passage of the Clean Air Act Amendments of 1990,[10] researchers agreed that transportation and air quality analysis must be integrated to assure progress toward the National Ambient Air Quality Standards (NAAQS) and revisions to transportation models were called for at the time (Burwell *et al.*, 1991, pp. 94–5). A general equilibrium analysis approach to the estimation of the social cost of environmental quality regulations has been developed (Hazilla and Kopp, 1990).

Environmental impact reviews of federal aid highway projects usually take years to complete. The General Accounting Office found that 70 percent of the projects that required a review under the National Environmental Policy Act (NEPA) required between two and six years to complete. They also found that more than 62 percent of the highway projects that required review under Section 404 of the Clean Water Act of 1977 took between four and eight years (USGAO, 1994).

Although efforts are underway to reduce such delays, there are a number of significant barriers to reducing delays in environmental approval. One of the most intransigent barriers is the inability to assess a given project's cumulative impacts on the environment. These impacts may become cumulative over time, through the interaction of noise, air quality, water quality, soil erosion, wildlife habitat, and so on.

A typical environmental impact statement considers a number of direct impacts on the transportation system and alternative road building scenarios (for example, projected daily traffic counts, or reduction in traffic across a bridge or highway segment). It also takes project costs into account. The estimated indirect effects include: the number of parks or recreation areas affected; and wildlife habitats, waterfowl's refuges, and wetlands crossed.

The current EPA model for estimation and prediction of air quality effects of mobile source emissions is MOBILE5a. California uses EMFAC and CAL3HQC software to estimate carbon oxide emissions. These models employ mobile source inventories which estimate and track vehicle–miles of travel. The effects of Transportation Control Measures (TCM) can be measured by these models. The outputs from these models are used in the State Implementation Plans (SIPs) as mandated by the Intermodal Surface Transportation Efficiency Act (ISTEA) of 1991,[11] the relevant implementing rules, and later in TEA21. A manual of regional transportation modeling practice for air quality analysis has been developed for the National Association of Regional Councils (Harvey and Deakin, 1993).

Equity outcomes
Methods for calculating the value of environmental externalities of infrastructure and regional development are being created and can be found in the literature (Quinet, 1993). Some have suggested that a "social impact analysis" be conducted in conjunction with an environmental impact analysis. Guidelines and principles of social impact analyses have been presented elsewhere (Interorganizational Committee, 1993).

All investments are known to have distributive or equity effects. Any assessment of ITS investment costs must consider the distribution of costs and benefits across groups and across geographic sub-regions within metropolitan areas. For example, an attempt to implement variable road pricing to reduce emissions or congestion is likely to disproportionately impact low-income persons, as well as inner-city versus edge city residents. Transportation investments, both road and transit, may also lead to regional access disparities, another form of equity effect. Therefore, estimating impacts and identifying workable offsetting measures must be considered.

Not all transportation system outcomes and objectives can be easily quantified, especially those regarding social or demographic equity. It appears that a combined approach, including both quantitative methods (for example, econometric models, multi-criteria assessment, or multi-objective programming), and qualitative approaches, is indicated. For example, debates about environmental and fuel source sustainability, or about the social equity effects of transportation system development, may be based on both quantitative and qualitative outcome measures.

The ITS investment evaluation problem, as explained here, supports the inclusion of a multi-criteria assessment approach to address these equitable and qualitative outcome measures. A thorough review of multiple criteria analysis with respect to physical planning is available (Nijkamp *et al.,* 1990). Multicriteria analysis has been used to explore future ITS applications in Europe (Reggiani *et al.*, 1994, pp. 16–23), where transport corridors are classified based upon the potential for introduction of advanced transport technologies. This potential is derived from the "compound value" of each corridor, in terms of its interdependencies and "socio-economic traffic" profile. Three main corridor evaluation criteria are used:

- socio-economic profile;
- transport supply profile;
- transport demand profile.

The family of multiple criteria methodologies ranges from simple checklists (for example, the environmental checklist) to highly sophisticated approaches such as multi-object programming. In all cases, the researcher is faced with a decision of whether to weight all criteria equally, or to apply different weights depending upon the importance of each criteria to the decision maker or policy maker.

Multiple criteria utility assessment is defined as a systematic means to assess the worth of complex alternatives. It provides a common scale for combining judgments on more than one dimension or for more than one criterion (Delp, 1977).

The general procedure starts with identification of the decision criteria used to assess each alternative. Then a utility function for each criterion is constructed (most preferred/least preferred) and utility functions for the criteria are constructed (0 percent to 100 percent). After testing criteria one at a time, one may determine the utility of each alternative. Then the decision maker must order (weight) the criteria, and by multiplication and aggregation, utilities may be calculated and ranked (ibid.). Haynes *et al.* (Chapter 8 in this volume) apply a multi-criteria utility model called PROSCAL to assess planners' views on an ITS bridge application assessment.

Transportation project selection with respect to equity, where there are only relative, or "fuzzy" objectives, may be improved through the use of fuzzy set theory. Transportation investment planning has multiple and often mutually conflicting social objectives. These objectives may include economic growth, cost, accessibility, mobility, quality of life, environmental concerns, integration of transport systems, national defense, and regional equity. A multi-objective integer programming solution for competing investments could be developed (Tzeng and Teng, 1993). Other uses of fuzzy set theory in this arena have been explored (Boettcher, 1994; Bellman and Zadeh, 1970).

Equity indicators include:

- affordability for the poor to use ITS equipment: cost as burden;
- accessibility for the poor to the new system: regional access disparities;
- usability for the disabled: percentage of people excluded.

Institutional outcomes

Any major transportation investment in metropolitan areas may have significant institutional effects. For example, the interstate highway system – in particular, the urban freeways built under this program – affected the spatial location and dispersion of housing and job markets, the structure of the motor carrier industry, and had serious competitive impacts on pre-freeway transport modes. In many cases, the interstate highway system – particularly urban interstates – shaped urban form, travel patterns, and way of life. ITS has a similar institutional transforming potential that should be considered as part of the assessment (Horan and Gifford, 1993).

Institutional effects include ownership variations, for example, some states own and control almost all their non-interstate road mileage, while in other states, localities (cities and counties) control a large share of the road mileage. Public–private partnerships have been suggested for transit operations (Cervero, 1994). Many states, though they do not yet measure institutional effects, recognize the importance of institutional dynamics (California Department of Transportation, 1993; Ohio Department of Transportation, 1993). The Federal Highway Administration has identified ways to evaluate institutional issues in ITS operational tests by looking at whether the test met its goals, and whether it generated any related projects (US Department of Transportation, 1992). The interested reader is referred to Stough and Rietveld (1997) for a more detailed examination of institutional issues in the transportation context.

Federal legislation over the past 10 years provides opportunities, incentives, and mandates for new institutional arrangements in policy making. ISTEA, as its successor, TEA21, altered a number of intergovernmental relationships (US Congress, 1991). Section 134 of ISTEA mandates that urbanized-area metropolitan planning organizations develop intermodal transportation plans (TIPs) for their respective metropolitan areas, with an eye toward mobility as well as air quality.[12] State congestion management plans and long-range transport plans (SIPs) are also included in this mandate. Under ISTEA, metropolitan area ITS issues such as clean air and water, energy conservation, and traffic congestion for the first time were dealt with at the metropolitan level by designated Metropolitan Planning Organizations (MPOs).

With ISTEA, and now TEA21, states are required to coordinate their statewide transportation planning efforts with metropolitan areas.[13] Today, states and MPOs must reconcile their transportation plans and programs for metropolitan areas.[14] MPOs, even though they may not participate financially in a state-led transportation infrastructure project, have the right of planning review.[15] The effects of these institutional arrangements should be included in efforts to evaluate ITS investments.

Institutional indicators are:

- ownership variations;
- intergovernmental relationships: ISTEA, TEA21, and MPOs;
- financial transfers and implications;
- liability and jurisdictional changes.

Methodological problems in pursuing the total-cost approach

A number of methodological and technical problems are inherent in the evaluation of transportation investments using the total-cost approach. The approach is characterized by a significant degree of complexity: these methodological and technical problems introduce the additional dimension of uncertainty. This uncertainty starts with questions about the necessary extent and intensity of assumptions which accompany the approach. There is also uncertainty regarding the proper form of the model to be employed. There is uncertainty introduced by the outcomes that require qualitative and normative assessments. For those relationships that can be quantitatively taken into a model framework (such as a regional econometric input–output model) there is uncertainty in estimating the second-order induced effects. Data limitations on the inter-temporal and inter-sectoral effects may compel the adoption of new assumptions and creative estimation techniques to address this complexity and uncertainty.

In using an input–output framework to estimate direct and indirect (induced) effects of a given transportation system investment, the first-order effects lead to second-order (induced) effects – the resulting changes in the system will feed back and promote changes in the first-order conditions. The direction of effects and interactions is two-way, which makes estimation difficult. The National Cooperative Highway Research Program has recognized this in a recent report that developed an analytical framework with supporting methods to evaluate the indirect effects of transportation projects.[16]

CONCLUSIONS AND FUTURE RESEARCH

Traditional input–output economic models can be used to estimate the economic (jobs, income, output, value-added) effects of ITS and other transportation investments. When these models are combined with an econometric component, it is possible to forecast economic effects and, therefore, to estimate the net present worth of proposed investments through discounting the estimated future stream of benefits. Furthermore, air pollution residual and fiscal sub-models can be constructed and linked directly to the input–output model so that these effects can also be estimated (Lakshmanan and Ratick, 1980). It should be noted that this approach could be used to estimate data for multi-criteria assessment.

The primary advantage of regional input–output econometric modeling is its ability to estimate many of the indirect effects of its investments and related transportation investments – this will contribute to a much more integrated and useful evaluation methodology. However, while this is a plausible and rigorous approach to the full social cost and benefit assessment problem, it would fall short of an easily implemented approach. Such an approach would be costly to implement and would not be totally capable of accounting for costs/benefits of externalities and especially difficult to measure components like the effects of institutional issues. Consequently, a more limited and heuristic multi-criteria approach is recommended for ITS project evaluation. Such approaches range from a simple checklist (see checklist table in the Appendix to this chapter) to the planning balance sheet approach (see Chapter 7).

APPENDIX: TRANSPORTATION ITS PROJECT IMPACT CHECKLIST

This basic checklist illustrates the dimensions of a full social costs and benefits multi-criteria approach to the evaluation of ITS projects.

Category	Indicator
Safety	Deaths
	Injuries
	Property damage
Congestion and delay	Peak-hour VMT
	Vehicle-hours of delay
	Peak average travel time
	Average length of queue
Capacity and mobility	Ease of personal mobility
	System accessibility
	Daily VMT
Cost	Construction
	Operations
	Maintenance
Economic outcomes	Gross regional product
	Employment
	Per capita income
	Competitiveness

Category	Indicator
Fiscal outcomes	Revenue enhancement
	Budgetary sufficiency
	Inter-governmental transfers
Environment outcomes	Ambient emissions levels
	Ground level ozone
	Soil erosion
	Wetlands and habitat
Equity outcomes	Affordability
	Regional access disparity
	Handicap access
Institutional outcomes	Ownership variation
	Inter-governmental frictions
	Liability and jurisdiction changes

NOTES

1. As noted in the Introduction, ITS will also be a significant component in the management of transport corridors (Beimborn and Horowitz, 1993; Marshment, 1993; Ryan and Emerson, 1986).
2. Road subsidies were estimated at $31 billion per year, and annual parking subsidies at $85 billion (Pucher and Hirschman, 1993).
3. US DOT, Federal Highway Administration, *Highway Statistics* (series, annual).
4. As described by Dr Daniel Rathbone, Editor, *Urban Mobility Monitor*. Interviews, September 1994.
5. See Callahan (1994) for an experimental approach to a regional production function.
6. See the estimation sub-routine in NVREIM for dynamic multipliers which can partially address this problem (Campbell and Stough, 1994).
7. For an analysis of the stimulative versus the substitutive effects of federal aid on local highway spending, see Walzer and Deller (1993).
8. P.L. 101–549 (1990).
9. P.L. 102–240 (1991).
10. P.L. 101–459 (November 15, 1990).
11. P.L. 102–240 (December 18, 1991).
12. P.L. 102–240, Sec. 1024 §134 (a).
13. P.L. 102–240, Sec. 1024 §135 (b).
14. P.L. 102–240, Sec. 1024 §135 (d)(1).
15. Federal Aid Highway Act of 1970; 1973 Federal Aid Highway Act.
16. The study, NCHRP Project 25-10, was conducted by Louis Berger and Associates (1994–5).

REFERENCES

Anderson, W.P., Kanaroglou, P.S. and Miller, E.J. (1994), "Integrated land use and transportation model for energy and environmental analysis: A report on design and implementation", Working Paper, McMaster University, Canada.

Arrow, K.J. and Kurz, M. (1970), *Public Investment, the Rate of Return, and Optimal Fiscal Policy*, Baltimore, MD: Johns Hopkins University Press.

Arthur, W.B. (1990), "Positive feedbacks in the economy", *Scientific American* February, 92–9.

Aschauer, D.A. (1989), "Is public expenditure productive?", *Journal of Monetary Economics* **23**(2), 177–200.

Beimborn, E. and Horowitz, A. (1993), *Measurement of Transit Benefits*, Springfield, VA: National Technical Information Service (Order No. 1PB 93-208122).

Bellman, R.E. and Zadeh, L.A. (1970), "Decision-making in a fuzzy environment", *Management Science* **17**(4).

Biehl, D. (1991), "The role of infrastructure in regional development", *European Research in Regional Science, Infrastructure and Regional Development* **1**, 9–35.

Boettcher, H. (1994), "Acceptance functions in public decision making: A new concept for regional infrastructure investments", presented at Western Regional Science Association Annual meeting.

Brand, D. (1991), "Research needs for analyzing the impacts of transportation options on urban form and the environment", in *Transportation, Urban Form, and the Environment* (Proceedings of a conference held December 9, 1990), Irvine, CA: Federal Highway Administration and Transportation Research Board (Washington, DC), pp. 101–16.

Burwell, D.G. (1991), "Energy and environmental research needs", in *Transportation, Urban Form, and the Environment* (Proceedings of a conference held December 9, 1990), Irvine, CA: Federal Highway Administration and Transportation Research Board (Washington, DC), pp. 81–100.

Button, K.J. (1982), *Transport Economics*, London: Heinemann.

California Department of Transportation, Division of New Technology, Materials and Research (1993), *Strategies for the Caltrans New Technology Development program (draft)*, Sacramento: Author.

Callahan, J. (1994), Increasing returns from knowledge infrastructure. Fairfax, VA: George Mason University, School of Public Policy (unpublished).

Campbell, H. and Stough, R.R. (1994), "Northern Virginia Regional Econometric Input–output Model (NVREIM)", George Mason University, Fairfax, VA.

Cervero, R. (1994), "Making transit work in the suburbs", presented at the 73rd Annual Meeting of the Transportation Research Board, Washington, DC, January.

Costa, J., Ellson R. and Martin, R. (1987), "Public capital regional output and development: Some empirical evidence", *Journal of Regional Science* **27**, 419–37.

Covil, J.L., Taylor, R.S. and Sexton, M.C. (1994), "Will multimodal planning result in multimodal plans?", presented at the 73rd Annual Meeting of the Transportation Research Board, Washington, DC, January (Paper 940205).

Deakin, E.A. (1991), "Jobs, housing and transportation: Theory and evidence on interactions between land use and transportation", in *Transportation, Urban Form, and the Environment* (Proceedings of a conference held December 9, 1990), Irvine, CA: Federal Highway Administration and Transportation Research Board (Washington, DC), pp. 25–42.

Delp, P. (1977), *Systems Tools for Project Planning*, Bloomington, IN: Indiana University.

DeLucchi, M. (1993), "Full social costs of transportation policy", Davis, CA: University of California at Davis (working paper).

Downs, A. (1962), *Stuck in Traffic: Coping with peak hour traffic congestion*, Brookings Institution and the Lincoln Institute of Land Policy, Washington, DC: Brookings Institution Press.

Downs, A. (1992), *Stuck in Traffic: Coping with peak hour traffic congestion*, Brookings Institution and the Lincoln Institute of Land Policy, Washington, DC: Brookings Institution Press.

Duffy-Deno, K.T. and Eberts, R.W. (1989), "Public infrastructure and regional economic development: A simultaneous approach", Working paper no. 8909, Cleveland: Federal Reserve Bank of Cleveland.

Eberts, R. (1986), "Estimating the contribution of urban public infrastructure to regional growth", Working Paper, Cleveland: Federal Reserve Bank of Cleveland.

Eberts, R.W. (1991), "Some empirical evidence on the linkage between public infrastructure and local economic development", in H.W. Herzog Jr and A.M. Schlottmann (eds), *Industry Location and Public Policy*, Knoxville: University of Tennessee Press, pp. 83–96.

Eisner, R. (1991), "Infrastructure and regional economic performance: Comment", Evanston, IL: Northwestern University, unpublished paper.

Feser, E. and Bergman, F.M. (2000), "National industry cluster templates: a framework for applied regional cluster analysis", *Regional Studies* 34(1), 1–19.

Gifford, J.L., Horan, T.A. and White, L.G. (1994), "Dynamics of policy change: Reflections on the 1991 federal transportation legislation", presented at the 73rd Annual Meeting of the Transportation Research Board, Washington, DC, January (Paper 940814).

Green, D.L., Jones, D.W. and Delucchi, M.A. (1997), *Measuring the Full Social Costs and Benefits of Transportation*, Heidleburg: Springer-Verlag.

Hartgen, D.T. and Krauss, R.T. (1993), "Resources vs. results: Comparative performance of state highway systems, 1984–1990", *Policy Studies Journal* 21(2), 357–74.

Harvey, G. and Deakin, E. (1993), *A Manual of Regional Transportation Modeling Practice for Air Quality Analysis*, Version 1.0, Washington, DC: National Association of Regional Councils.

Hazilla, M. and Kopp, R.J. (1990), "Social cost of environmental quality regulations: A general equilibrium analysis", *Journal of Political Economy* 98(4), 853–73.

Hewings, G.J.D. and Jensen, R.C. (1986), "Regional, interregional and multiregional input–output analysis", in P. Nijkamp (ed.), *Handbook of Regional and Urban Economics*, vol. 1, Amsterdam: North-Holland.

Hicks, J.R. (1946), *Value and Capital: An inquiry into some fundamental principles of economic theory*, 2nd edn, Oxford: Clarendon Press, 1974.

Hill, E. and Brennan, J.F. (2000), "A methodology for identifying the drivers of industrial clusters: the foundation of regional competitive advantage", *Economic Development Quarterly*, 14(1), 65–96.

Holtz-Eakin, D. (1988), "Private output, government capital, and the infrastructure crisis", Discussion paper no. 394, New York: Columbia University.

Holtz-Eakin, D. (1992), "Public sector capital and the productivity puzzle", Working Paper no. 4144, Washington, DC: National Bureau of Economic Research.

Horan, T.A. (1992), "Evaluating IVHS: Key issues in institutional and environmental assessments of IVHS technologies", in J. Gifford, T. Horan and D. Sperling (eds), *Transportation, Information Technology and Public Policy*, Fairfax, VA: George Mason University, pp. 191–209.

Horan, T. (ed.) (1993), *National IVHS and Air Quality Workshop* (Proceedings), Diamond Bar, CA: George Mason University (Fairfax, VA).

Horan, T.A. and Gifford, J.L. (1993), "New dimensions in infrastructure evaluation: The case of non-technical issues in intelligent vehicle-highway systems", *Policy Studies Journal* 21(2), 347–56.

Hotchkiss, W.E. (1977), "Cost-benefit analysis", in D.A. Hensher (ed.), *Urban Transport Economics*, Cambridge: Cambridge University Press, pp. 44–54.

Interorganizational Committee on Guidelines and Principles (1993), *Guidelines and Principles for Social Impact Analysis*, Belhaven, NC: International Association for Impact Assessment.

Investment in Transport Infrastructure in the 1980s (1992), Washington, DC: Organization for Economic Cooperation and Development (OECD).

Jones, D.W. (1995), "Transportation and overall economic performance: Views on recent research from several perspectives", in *Transportation Statistics Annual Report 1995*, Washington, DC: Bureau of Transportation Statistics, US Department of Transportation.

Kuhn, T.E. (1965), "The economics of transportation planning in urban areas", in *Transportation Economics*, New York: National Bureau of Economic Research, Columbia University, pp. 297–325.

Lakshmanan, T.R. and Nijkamp, P. (eds) (1980), *Economic–Environmental–Energy Interactions*, Boston, MA: Martinus Nijhoff.

Lakshmanan, T.R. and Ratick, S. (1980), "Integrated models for economic energy: Environmental impact analysis", in T.R. Lakshmanan and P. Nijkamp (eds), *Economic–Environmental–Energy Interactions*, Boston, MA: Martinus Nijhoff.

Leontief, W. (1966), *Input–Output Economics*, New York: Oxford University Press.

Lorenzini, S. (1994), "The effects of a high-speed train tunnel: A case study on Florence", presented at *Transport Econometrics: Effects of new major infrastructures*, Calais, France: Applied Econometrics Association.

MacKenzie, J.J. (1994), "The going rate: What it really costs to drive", presented at the 73rd Annual Meeting of the Transportation Research Board, Washington, DC (Paper 940986).

Marshment, R.S. (1993), "Establishing national priorities for rail transit investments", *Policy Studies Journal* 21(2), 338–46.

Masser, I., Svidén, O. and Wegener, M. (1993), "Transport planning for equity and sustainability", *Transportation Planning and Technology* 17(4), 319–30.

Mera, K. (1973), "Regional productions and social overhead capital: An analysis of the Japanese case", *Regional and Urban Economics* 3, 157–85.

Mishan, E.J. (1988), *Cost-benefit Analysis*, London: Unwin-Hyman.

Mohring, H. and Anderson, D. (1994), *Estimating Congestion Costs and Optimal Congestion Charges for Large Urban Areas*, St. Paul, MN: Metropolitan Council.

Mudge, R.R. (1992), "Approaches to the economic evaluation of IVHS/ITS technology", in J. Gifford, T. Horan and D. Sperling (eds), *Transportation, Information Technology and Public Policy*, Fairfax, VA: George Mason University, pp. 97–126.

Mudge, R.R. (1994), "Full costs of transportation", presented at the 73rd Annual Meeting of the Transportation Research Board, Washington, DC.

Munnell, A.H. (1990a), "Is there a shortfall in public capital investment: An overview", in A.H. Munnell (ed.), *Is There a Shortfall in Public Capital Investment?*, Boston, MA: Federal Reserve Bank of Boston, pp. 1–20.

Munnell, A.H. (1990b), "Why has productivity growth declined? Productivity and public investment", *New England Economic Review* Jan.–Feb., 3–22.

Murphy, J.J. and Delucchi, M.A. (1998), "A review of the literature on the social cost of motor vehicle use in the US", *Journal of Transport Statistics*, January, pp. 15-42.

Nijkamp, P., Rietveld, P. and Voogd, H. (1990), *Multiple Criteria Analysis in Physical Planning*, Amsterdam: North-Holland.

Ohio Department of Transportation (1993), *Access Ohio: Ohio Multi-modal State Transportation Plan to the Year 2020*, Columbus, OH: Author.

Pisarski, A.E. (1991), "Overview", in *Transportation, Urban Form, and the Environment* (Proceedings of a conference held December 9, 1990), Irvine, CA: Federal Highway Administration and Transportation Research Board (Washington, DC), pp. 3–10.

Porter, M.E. (2000), "Location, competition and economic development: local clusters in a global economy", *Economic Development Quarterly* 14(1), 15–34.

Prud'homme, R. (1993), "Assessing the role of infrastructure in France by means of regionally estimated production functions", presented at *Infrastructure, Economic Growth and Regional Development: The case of industrialized high-income countries*, Jönköping, Sweden.

Pucher, J. and Hirschman, I. (1993), "Public transport in the United States: Recent developments and policy perspectives", *Public Transport International* 41(3), 9.

Quinet, E. (1993), "Valuation of environmental externalities: Some recent results", presented at *Infrastructure, Economic Growth and Regional Development: The case of industrialized high-income countries*, Jönköping, Sweden.

Reggiani, A., Lampugnani, G., Nijkamp, P. and Pepping, G. (1994), "Towards a typology of European interurban transport corridors for advanced transport telematics (ATT) applications", presented at George Mason University, School of Public Policy, Fairfax, VA.

Roberts, B.H. and Stimson, R.J. (1998), "Multi-sectoral qualitative analysis: assessing the competitiveness of regions and developing strategies for economic development", *Annals of Regional Science* 32(2), 15.

Ryan, J.M. and Emerson, D.J. (1986), *Procedures and Technical Methods for Transit Project Planning*, Washington, DC: Federal Transit Administration.

Rycroft, R.W. and Kash, D.F. (1999), *The Complexity Challenge: Technological innovation for the 21st century*, London and New York: Pinter.

Schulz, D.F. (1991), "Decision makers need help", in *Transportation, Urban Form, and the Environment* (Proceedings of a conference held December 9, 1990), Irvine, CA: Federal Highway Administration and Transportation Research Board (Washington, DC), pp. 11–24.

Stough, R.R. and Haynes, K.E. (1997), "Megaproject impact assessment", in M. Chatterji (ed.), *Regional Science Futures*, London: Macmillan.

Stough, R.R. and Maggio, M.E. (1994), "Evaluating IVHS/ITS transportation infrastructure in a metropolitan area", presented at 1994 Korea–USA Symposium on IVHS and GIS-T, Seoul, June.

Stough, R.R. and Rietveld, P. (1997), "Institutional issues in transportation systems", *Journal of Transportation Geography* 5(3), 207–14.

Stough, R.R., Stimson, R.J. and Roberts, B.H. (2000), "Merging quantitative and expert response data in setting regional economic development policy: methodology and application", Working paper, Fairfax, VA: Institute of Public Policy, George Mason University.

Tatom, J.A. (1993), "The spurious effect of public capital formation on private sector productivity", *Policy Studies Journal* **21**(2), 391–5.

Tzeng, G.H. and Teng, J.Y. (1993), "Transportation investment project selection with fuzzy multiobjectives", *Transportation Planning and Technology* **17**(4), 91–112.

US Congress (1991), *Intermodal Surface Transportation Efficiency Act of 1991* (Public Law 102-240), Washington, DC: US Government Printing Office.

US Department of Transportation, Federal Highway Administration (1987), *Highway Statistics 1987*, Washington, DC: US Government Printing Office.

US Department of Transportation, Federal Highway Administration (1992), *Public and Private Sector Roles in IVHS Deployment: Final Report*. Summary of Seminar Proceedings, Washington, DC: US Government Printing Office.

US General Accounting Office (1994), *Highway Planning: Agencies Are Attempting to Expedite Environmental Reviews, but Barriers Remain*, Letter Report, 08/02/94, GAO/RCED-94-211. Washington, DC.

University of Florida (1990–2000), *McTrans Catalog*, Gainsville, FL : Transportation Research Center, Center for Microcomputers in Transportation.

Waldrop, M.M. (1992), *Complexity, the Emerging Science at the Edge of Order and Chaos*, New York: Simon and Schuster.

Walzer, N. and Deller, S.C. (1993), "Federal aid and rural county highway spending: a review of the 1980s", *Policy Studies Journal* **21**(2), 309–24.

3. Electronic Toll Collection (ETC) in the Dulles Corridor

Roberto Mazzoleni

INTRODUCTION

This chapter evaluates the process and outcomes of the implementation of an electronic toll collection (ETC) system on two major tolled highways in the Northern Virginia part of the National Capital region – the Dulles Toll Road (DTR) and the Dulles Greenway (DG). The methodology for evaluation is based on the checklist approach (see Appendix 3 to this chapter) to multi-attribute assessment developed by Stough *et al.* (1994) and further elaborated in Chapter 2. The general justification for this approach is particularly relevant for the ETC applications examined here. In fact, summarizing the outcomes of the projects in a single measure (as typically done in cost-benefit analysis) would be difficult because of the multiplicity and diversity of the projects' goals as well as for the diversity of social actors who are impacted by the projects' benefits and costs.

It is also argued that only a subset of the many possible outcomes identified in the checklist occur at the implementation site of this study. Because ETC projects operate on particular links of a complex road network structure, their realization affects connected nodes and links of that network, thus contributing to the overall effectiveness of the transportation system. The ETC projects' external effects are not solely confined to the functionality of the road network because of the ensuing influences on inter-modal choice, residential and business location, and on related policy initiatives and inter-jurisdictional co-operation/conflict. Accordingly, the evaluation of the project occurs at different levels of spatial, economic and social aggregation.

The first part of the chapter provides an introduction to the relationship between congestion and tolling, illustrating the potential benefits to be gained from the introduction of ETC systems. The state of the transportation network in the Dulles Corridor is then examined followed by the introduction of the checklist methodology. Next the ETC projects are evaluated. The effects of ETC on congestion in the Dulles Corridor are examined in greater detail in the next section of the chapter. A discussion of the lessons that can be drawn from the analysis concludes the chapter.

TOLLING, CONGESTION, AND CAPACITY EXPANSION: GOALS OF ETC

Congestion of a roadway network has been an increasingly common phenomenon, particularly in metropolitan areas. The emergence of congestion is determined by a mismatch between network capacity and travel demand: as the number of vehicles using the network within a given time period increases, network functionality decreases. Traffic flows per hour decrease, raising transportation costs for users – due to increased travel time, higher fuel consumption, and wear and tear of vehicle components – as well as causing incremental costs for non-motorists in the form of increased emissions and reduced mobility for pedestrians and cyclists.

The emergence of congestion has been often associated with the external effects of individual motorists' travel decisions. Accordingly, it is argued that each driver's decision to use a particular link of the roadway network neglects costs that such decisions impose on others – motorists and non-motorists. As the level of use of a network link approaches its capacity, these external costs become significant and lead to the well-known outcome of traffic volumes that are in excess of socially desirable levels. Efforts to reduce congestion typically involve expansion of network capacity and forms of traffic management aimed at redistributing travel demand across network links and shifting the time patterns of travel demand (Mohring, 1999).

The redistribution approach to congestion reduction has become increasingly difficult because of the rising costs of right-of-way acquisitions in densely populated areas. Furthermore, empirical studies indicate that additional open-access capacity in a network is likely to produce changes in travel demand that over a period of time will reproduce the conditions of congested traffic. This is the basis for the black-hole theory of highway investment (Plane, 1995).

Tolling (or road pricing) has often been advocated in order to force individual motorists to internalize the external cost of their travel decisions on other drivers – and thereby reduce congestion (Pigou, 1920; Vickrey, 1969). However, practical implementation of road pricing schemes has traditionally deviated from the theoretical ideal. When tolls are levied on users of individual network links, the effect may be to redistribute travel demand across routes and, thus, shift congestion to other parts of the network. This effect becomes particularly undesirable when tolls do not vary by time of day or congestion levels, so that they cause inefficiently low traffic volumes on the tolled facility during off-peak hours while increasing traffic on non-tolled network links. The use of variable tolls (congestion pricing) has not been wide-spread in the US because tolling has been typically used for revenue generation rather than travel demand management (Gómez-Ibáñez and Meyer, 1993).

Furthermore, tolling's potential congestion mitigation benefits are partly offset by additional social costs, borne partly by highway operators – for building, maintaining and operating toll plazas, by motorists – whose travel time will increase, and by local residents – because of increased emissions from vehicles queuing at toll plazas.

These relationships between highway capacity, congestion and tolling, make it possible to identify the potential social benefits of the introduction of ETC systems, a recent technological innovation that enables highway operators to collect tolls from road users by electronic means. No physical exchange of money takes place at the toll collection point: the transaction is instead completed through the exchange of messages between a roadside transmitter and a vehicle-installed transponder. These messages enable the system to identify a specific vehicle and to charge the appropriate fees to the user's account.

The implementation of ETC on tolled facilities has the potential to generate important social benefits. First, ETC should reduce toll collection costs for highway operators relative to conventional manned or unmanned systems. Second, by eliminating congestion and queuing at the toll plazas, ETC may reduce travel costs for drivers and the environmental costs of congestion for neighboring communities. Third, the benefits of reduced congestion at toll plazas may spill over to other tolled and untolled links in the local transportation network. In addition to their primary congestion-relief function, ETC installations represent also an alternative or a complement to physical infrastructure investment projects where the primary goals are to create highway capacity to support or induce economic development and to support mobility.

These objectives, albeit in different degrees, play a role in the deployment of ETC systems at the DG and the DTR.[1] These two tolled facilities are located in the Dulles Corridor, one of the fastest growing regions in the country, extending in NW/SE direction around the Dulles International Airport in the Washington, DC area. The economic development of the last 20 years and the concurrent changes in the location of business activities and residential communities have contributed to the creation of a severe traffic congestion problem in the Washington, DC area, second only to Los Angeles in the nation.

TRAVEL DEMAND AND HIGHWAY CAPACITY IN THE DULLES CORRIDOR

The DTR and the DG together provide a direct highway link between the Capital Beltway and the Town of Leesburg through the Dulles Corridor. The economic and demographic characteristics of this area are far from homogeneous. A rough partition can be drawn between the districts east of Route 28 and Dulles International Airport (largely within the Fairfax County boundary) and those west of the Route and the airport (Loudoun County). The eastern section of the Corridor is one of the most economically successful regions in Northern Virginia and throughout the nation. The western section of the Corridor has not yet experienced similar economic development and it is still predominantly characterized by agricultural activity, forested and open land, as well as some residential development, although recently there has been accelerated residential development. This heterogeneity is reflected in the variety of expectations that stakeholders in the two ETC projects hold concerning their social, demographic and economic impacts.

The DTR, which has been in operation since 1984, has been a central element of the transportation network supporting rapid economic and demographic growth of the eastern section of the Dulles Corridor during the last two decades.[2] Such growth is expected to continue for the next 15 years (Table 3.1). The associated rise in traffic volumes will continue to put the local transportation network under severe strain, as in the past.

Table 3.1   Population and employment trends in the Dulles Corridor

	1990		2000		2010	
	Pop.	Empl.	Pop.	Empl.	Pop.	Empl.
Fairfax County Districts	262 403	187 222	321 548	251 700	362 751	318 320
Loudoun County Districts	63 575	27 371	95 602	43 197	135 947	62 003

Source: Metropolitan Washington Council of Governments (1998)

Since its opening in 1984, traffic volumes on the DTR have recurrently exceeded the planners' expectations. Based on the forecast that traffic would rise from about 29 000 vehicles per day in 1984 to 49 000 in the year 2000, the original design was assumed to provide sufficient capacity for about 20 years.[3] By the year 1986–7 the traffic volumes were still within the design service volume (DSV), but the initial projections underestimated the pace of traffic growth quite significantly.[4] During the next five years, the traffic increased quite rapidly, particularly in the western segments of the highway (Table 3.2). More recent forecasts estimate that traffic will rise to an average of 130 000 cars daily by 2010 with a deterioration of service level from E to F.[5] In comparison the DG, which opened only six months ago, has an average traffic volume of 10 500 vehicles per day and operates well below design capacity, although by 2000, some four years later, the vehicle volume per day had increased to 47 000 per day (weekdays).

Table 3.2   Annual average daily traffic on the Dulles toll road

	1986–7	1991	1993 (6 lanes)	2010
Rt28 to Rt657	13 380	27 000	35 000	69 600
Rt657 to Rt602	13 380	36 000	60 000	92 600–109 000
Rt602 to Rt674	22 830	59 000	63 000	123 000
Rt674 to Rt7	54 575	71 000	75 000	132 400–135 400
Rt7 to Rt495	45 790	73 000	77 000	108 800
Rt495 to Rt123	34 320	52 000	57 000	90 245
Rt123 to Rt66	34 320	47 000	49 000	90 245

Source: Virginia Department of Transportation (1989)

Congestion on the DTR has so far been dealt with through additional physical infrastructure, including a third lane and larger numbers of lanes at the toll plaza, but the relief brought about by these measures has proved temporary and generally insufficient. Further expansions of the highway capacity by similar methods face special constraints.[6] After the projected expansion to four lanes, there will not be much room for further expansions, thus requiring a novel and experimental light-rail/bus design or expensive subterranean or elevated heavier rail. In light of these constraints, the area constitutes an ideal environment for the deployment of ITS technologies.

However, the ITS approach has not been pursued to the exclusion of more conventional investment projects. In particular, the DG between Leesburg and Dulles International Airport has recently added a new link to the transportation network. Undertaken by Toll Road Investors Partnership II (TRIP II), a private sector entity, the investment was allowed by the Virginia Commonwealth that in 1988 passed legislation determining the regulatory framework for private investment in highway infrastructure.[7]

TRIP II's profit expectations are based on the belief that the DG will attract motorists currently using Routes 7, 28, and 50, connecting Loudoun County and more remote western districts with the closer sub-regions of the metropolitan area. Given the congestion levels on those expressways, the DG should produce economic value to the users that the highway operator can partly appropriate by imposing tolls. In addition to the diversion of traffic from other routes, the DG is also expected to serve and benefit from the foreseen increase in business and industrial activities on the areas lying west of the Dulles Airport.[8]

EVALUATION PROBLEMS AND CHECKLIST METHODOLOGY

The problems faced in the *ex ante* evaluation of the impacts of ETC are manifold. In addition to the direct effects of ETC (for example, traffic flow across toll plazas, local level of carbon monoxide emissions, and highway operator's costs and revenues), one should gauge the indirect and higher-order effects on the regional transportation network, the pattern of land use, and economic development (residential or business). Invoking the traditional tools of cost-benefit analysis will not be satisfactory for a variety of reasons.

For one thing the conversion into monetary terms of the welfare consequences of a project is a task whose difficulty ranges widely over the dimensions along which the outcomes occur. Furthermore, the viewpoint is gaining increased acceptance that the relationship between transportation system effectiveness and economic development is not uniquely defined, so that investments in the former may or may not stimulate the latter. When it does, the dynamics of development are characterized by increasing returns from positive feedbacks (see Arthur, 1990).

Considering the novelty of the technological and institutional aspects of ITS, it is also important to evaluate the process of implementation. The overall social cost of these projects is influenced by the procedures through which their development, deployment and management take place. While the novelty of the technologies (at least with respect to the techno-organizational competencies of the relevant public agencies) calls for establishing new procedures, the process through which the latter will take shape is likely to be one of trial and error. Analyzing the early experiences with the deployment of ITS technologies can diffuse the benefits of learning beyond the boundaries of particular projects.

Although the methodological problems outlined above have yet to be solved in a comprehensive manner, it is legitimate to start tackling them by developing and applying evaluative frameworks such as the checklist approach (Stough *et al.*, 1994). At the acceptable cost of some analytical and quantitative rigor, the checklist approach provides a comprehensive and flexible tool for examining the dimensions along which outcomes of interest to different stakeholders occur.

SUMMARY EVALUATION AND IDENTIFICATION OF MAJOR EVALUATION ISSUES

The evaluation of the ETC systems' deployment along several dimensions is summarized in the checklist table in the Appendix to this chapter. Here the discussion focuses on the general relevance of the more important categories of goals to the deployment of ETC and provides an evaluation of the expected outcomes.

Congestion and Delay

The most important effects of ETC are expected in the area of congestion and delay. The technical performance characteristics of ETC (expressed as the flow capacity of toll lanes) provide the most impressive evidence of the potential benefits from its application. However, the traffic conditions on the entire Dulles Corridor road network reflect the presence of several bottlenecks, including but not limited to the toll plazas. It is well known from bottleneck analysis that in the context of a network the effect of increasing the capacity of any particular bottleneck cannot be determined independently from its position in the network itself (Braess, 1968; Arnott *et al.*, 1993). Thus, although the deployment of ETC increases the flow capacity of toll plazas located at different points in the highway network, its effects on the actual traffic flows and travel costs in the region are uncertain.

Because of the physical location of many toll plazas at the on- and off-ramps of the DTR, traffic flows at the plazas are highly sensitive to both upstream and downstream traffic conditions. The proximity of toll plazas to other bottlenecks (for example, single-lane ramp junctions and signalized intersections) may place upper limits to the traffic flows that can be realized across them, even after the installation of ETC. Indeed, there is evidence that at peak-hour the congestion of the DTR ramps and mainline may render the actual benefits of ETC significantly less than the potential ones. However, this dampening effect will fade as travel demand decreases (as in the off-peak hours), consequently making the toll collection point the most severe bottleneck in the network.

In the case of the DG, current levels of travel demand are such that ETC will not be called upon to ease congestion problems. It is difficult to estimate what the impact of ETC *per se* will be on the route choice of commuters traveling between Leesburg and the inner suburbs of Washington. At the current levels of ridership the marginal impact of the ETC is likely to be minimal. As travel demand in the Loudoun County portion of the Dulles Corridor continues to rise, the new technology will produce higher benefits for users.

Capacity and Mobility

In general, mobility will be enhanced thanks to the shorter trip duration afforded by ETC. This will be particularly true for local trips that make use of the DG and the DTR. These benefits will be significantly reduced during the peak-hour when the network is operating at service level E or F. No significant effects can be expected on passenger ridership under the current uniform toll pricing.

Safety and Accidents

The DTR has a positive safety record, with very few accidents occurring at the toll plazas. Speed limit enforcement at the toll plazas and use of dedicated ETC lanes should keep accidents and injuries to the same low levels. However, the use of ETC on on-ramps may cause increased merge flows at the ramp–freeway junctions, which are known to propagate congestion throughout the mainline lanes.

Cost Relative to Other Options

The cost of additional toll lane capacity is lower with ETC than with other options such as new toll plaza lanes. Operating and maintenance costs will be also lower, based on reports from other ETC installations in the country (Pietrzyk and Mierzejewski, 1993, p. 23). However, per unit costs will vary according to the utilization of available capacity. In the long run, these costs will be reduced by the increasing diffusion of ETC among users.

Financial and Fiscal Outcomes

Under current pricing policies, ETC's ability to enhance revenues will be proportional to the number of vehicles using the facilities. Consequently, the reduced costs of toll collection will result in better operating margins, so that the investment will likely have a short pay-back period, ultimately generating profits for the DG and investment funds for VDOT.

Environmental Outcomes

The impact of ETC systems on emissions level is uncertain. Although reduced congestion at the toll plazas may reduce emissions, increased traffic volumes may offset these benefits. If congestion is transferred to downstream ramps, emission levels may simply be redistributed.

Economic Outcomes

The implications of ETC adoption for the regional economy is a delicate issue. It is common to analyze the economic consequences of investment in transportation infrastructure by creating a divide between the construction period and the post-construction. This mirrors fairly closely the distinction between first-order, or direct, effects of the investment and induced ones.

The direct effects of investment in ETC systems are likely to be somewhat anomalous relative to those that could be attributed to a more conventional project (for example, the laying out of a new highway lane), as the kind of work involved in ETC deployment is much different. Highly specialized equipment procured from high-technology companies that do not reside in Northern Virginia limits the contract work available to local companies. Consequently, only a small portion of the investment expenditure is captured in the regional economy. On the other hand, the disruption caused by the project to business activities and traffic conditions is minimal.

With respect to the induced effects, existing research on the relationship between the efficiency (or some appropriately defined characteristic) of the transportation infrastructure and regional economic growth does not indicate unequivocally a positive correlation between the two.[9] Furthermore, most studies focus on the effects from the addition of network links rather than expansion of existing link capacity in heavily congested transportation networks. Consequently, their results are not generalizable to the case considered here.

Equity Outcomes

Although the DTR is not planning to raise the toll after ETC implementation, the DG charges a much higher toll than the DTR. This may limit the affordability of the highway and lead to the segregation of commuting patterns by income level, with the toll facility mostly serving the wealthier fraction of the population or those who find reductions in travel time valuable. However, those not choosing to pay the toll have access to other non-tolled routes that parallel the DG.

Projected increases in land values around the DG suggest that there may be negative outcomes on regional access disparity. However, it is difficult to argue that ETC system implementation will be directly responsible for rising land values and subsequent reduced regional access. Proximity of the west part of the region to the Dulles International Airport is probably a greater driver of land values.

Institutional Outcomes

The implementation of ETC technology should provide increased flexibility in the configuration of tolls and their use for congestion pricing. Although neither of the two tolling agencies appears to be preparing for the use of congestion pricing, the ETC will increase the range of pricing policies which they can pursue.

The implementation of the ETC systems has not required significant intergovernmental bargaining at the planning and deployment stage. However, the early inclusion of local authorities, business and resident communities in these stages may have generated a diffuse commitment to solving future problems in a cooperative mode. In contrast, the adjacent location of the two highways has enhanced the importance of ETC system compatibility and, therefore, of cooperation between the projects' sponsors – VDOT and TRIP II. While the goal of technical system compatibility was ultimately achieved, the relationship between the two highway operators has failed to perform as intended.[10]

MAJOR EVALUATION ISSUES

Highway Configuration Around Toll Plazas and Impact of ETC on Local Traffic Flows

One of the primary goals of ETC implementation is to increase the throughput capacity of a tolled facility. This objective is particularly important when toll plazas constitute a binding bottleneck on the traffic flows in a link or local network.[11] The problem of evaluating the effect on travel costs of changes in the capacity of a network link cannot be solved analytically in the case of moderately complex networks.[12] Existing models of simple networks indicate that, under certain conditions, increasing capacity of a network link may increase travel costs (Braess, 1968; Arnott *et al.*, 1993). In light of these difficulties, this discussion focuses simply on micro-level details of the ETC installation sites in order to gain some insight into their effects on local traffic capacity.

On the DTR and the DG, tolls are collected at various ramp junctions as well as at the mainline terminus. The general pattern on the DTR is to have toll plazas on the eastbound on-ramps and the westbound off-ramps. Toll lanes at on-ramp locations operate as a ceiling on the rate at which vehicles can enter the main highway segments. The introduction of ETC will in principle increase the maximum rate at which vehicles can enter access ramps leading to the DTR's mainline. Conversely, the toll lanes on off-ramps regulate the rate at which vehicles can clear the off-ramp lanes and access other roadways.

The introduction of ETC will increase this rate. It is necessary then to assess the possibility, as well as the desirability, of increasing the rate of merge flow at ramp junctions in light of the current traffic conditions.[13] A heuristic approach to this problem can be framed in terms of bottleneck analysis. In particular, it can be of interest to verify whether toll lanes constitute the most severe bottlenecks.

On-ramps

For the on-ramps vehicles (a) access the ramps from another roadway; (b) travel through the toll plaza; (c) proceed to the ramp junction and merge into mainline traffic (see Figure 3.1). Flow conditions on the on-ramp depend on the relationship between the volume and capacity of roadway segments:

- lanes leading to the toll plaza;
- toll plaza;
- ramp junction.

Congestion will occur at any of these points if demand exceeds a facility's capacity. Note that the capacity of each upstream link determines the maximum rate at which vehicles will enter the next link. Thus, if the capacity of (a) is less than that of (b), segment (b) will not be congested as the rate of flow is constrained by (a). In this case increasing the capacity of (b) will not have any effect on travel time: bottleneck (b) is hidden downstream. Also, when the capacity of (c) is less than that of (b), congestion may occur at (c) and propagate upstream. In this case too, increasing the capacity of (b) is not useful since (b) is a hidden upstream bottleneck.

For the off-ramps, the vehicles (a) diverge from the mainline flow by entering the ramp lanes; (b) go through the toll plaza; and (c) proceed to merge into the traffic along the intersecting roadway (see Figure 3.1). The analysis by travel segments carries through to this case.

This taxonomy indicates under just what conditions the increased capacity at toll plazas through the introduction of ETC (in dedicated or mixed-mode lanes) will serve to reduce travel time (see summary in Table 3.3). The design capacity of dedicated ETC lanes is believed to be 1300 vehicles per hour per lane (vphpl) or higher (Pietrzyk and Mierzejewski, 1993). In comparison, manual toll collection systems can clear up to 350 vphpl. On-ramp and off-ramp installations are likely to employ ETC in a mixed mode, so that the capacity should be 700 vphpl (ibid.). However, the extent to which the increased capacity of toll lanes will result in shorter trip duration hinges on time- and site-specific patterns of travel demand.

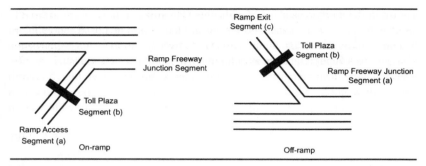

Figure 3.1   Configuration of tolled ramps on the Dulles toll road

The characteristics of travel demand on the DTR are quite typical, with a morning and evening rush hour serving commuter traffic traveling eastbound to work and westbound back to their residences. The average weekday daily traffic at the DTR's mainline plaza in October 1994 was set by the VDOT at 92 234 (both directions). Analysis of the hourly traffic variations at the same location reveals that the peak hour is 7 a.m. to 8 a.m. in the eastbound direction (10.8 percent of daily eastbound traffic) and 5 p.m. to 6 p.m. in the westbound direction (10.8 percent). Around 30 percent of the daily traffic transits across the mainline plaza in the three-hour periods 6 a.m. to 9 a.m. (eastbound direction) and 4 p.m. to 7 p.m. (westbound direction).[14] Although there are no official data, it is estimated that about 40 percent of the vehicles using the DTR travel its entire length.[15] Thus, a large share of the vehicles that cross the mainline plaza access the DTR from ramps located at intermediate points.

A study by JHK & Associates Inc. (May 1995) provides data on the peak-hour level of merge/diverge flow rates at the ramp junctions of the DTR. On the basis of traffic flow analysis the study suggests that peak-hour levels of service are at E or F at various locations (see Table 3.4). In the case of tolled on-ramps a level of service E or F at the ramp–freeway junction indicates congestion downstream from the toll plazas: a situation equivalent to the two configurations in Table 3.3 wherein segment (c) may be congested and the toll plaza is not a binding bottleneck. For these ramps, increasing the capacity of the toll plaza will have limited or no effect on peak-hour travel time and queue length, as the increased traffic flow will only add to congestion downstream. For the off-ramps, the analysis would require information on the traffic conditions downstream from the toll plazas: information which is not available. If these constituted the operating bottleneck, an increase in their capacity would speed up the outflow of cars from highway off-ramps into connected roadway links. However, if the ramp–freeway junction was processing traffic at a rate less than the capacity of the toll plazas, then increasing the plaza's capacity would not have significant effects.

The analysis can be adapted to eastbound and westbound traffic at the main-line toll plaza. For eastbound traffic, the flow downstream from the toll collection points feeds into connected roadways such as the Capital Beltway, Route 123 and Chain Bridge Road. In this respect, it is worth noticing that the

Table 3.3   Bottleneck analysis for tolled ramp configuration

Ranking by capacity	Bottlenecks[a]		Congested segments	Effect of increasing capacity of (b)
	Operating	Hidden		
a>b>c	c	b	c, b, a	Limited
a>c>b	b	c	b, a	Reduce congestion
b>a>c	c	–	c, b, a	None
b>c>a	–	–	–	None
c>a>b	b	–	b, a	Reduce congestion

Note: [a]   Assuming that traffic flow is at least equal to capacity of segment (a).

Table 3.4   Peak-hour traffic on tolled ramp junctions on the Dulles toll road

Ramp junction	On/Off	Direction	Merge flow rate		Level of service	
			a.m. peak	p.m. peak	a.m.	p.m.
Sully Road	On	E	990	420	B	B
	Off	W	920	2050	C B	F B
Centreville Road	On	E	1230	980	D C	C B
	Off	W	630	1580	B B	D C
Fairfax County	On	E	1140	780	F D	C B
Parkway	Off	W	530	1830	C B	F E
Reston Avenue	On	E	890	640	D E	C C
	Off	W	820	820	C C	F E
Wiehle Avenue	On	E	790	530	F E	C C
	Off	W	470	810	C C	E F
Hunter Mill Road	On	E	820	520	F F	C C
	Off	W	390	670	C C	E F
Leesburg Pike	Off	E	1310	810	F E	C B
Spring Hill Road	Off	E	800	220		
	On	E	230	1180	E D	F C
	Off	W	510	290	B C	B D
	On	E	80	750	C	D

Source: JHK & Associates (1995)

peak-hour level of service at the junction with the Beltway is E on the main-line and F on the ramp–freeway junction. The Beltway junction serves about 50 percent of the eastbound/westbound peak-hour traffic (ranging from 2500 to 3100 vehicles). Although the effects of ETC on congestion at the eastbound DTR mainline plaza are likely to be minimal, the situation is more favorable for westbound traffic as there is no recurrent congestion downstream.

Although the configuration of toll plazas on the DG mirrors, quite closely, that of the DTR, traffic flows on the DG are well below capacity; so that, while the toll plazas constitute the most severe bottlenecks, they do not result in queues and congestion.[16] For the present, the benefits of ETC implementation are somewhat limited due to low travel demand. However, in a longer perspective, the growth of travel demand will be a crucial issue.

The foregoing analysis indicates that travel time savings will be greater in the absence of downstream and upstream bottlenecks, conditions that during peak-hours are common only to the westbound and eastbound mainline toll plaza and a few ramps. During off-peak hours, traffic flows decrease consid-erably, which suggests that the current toll collection system is operating be-low capacity. If this is the case, the travel time advantages of ETC are rela-tively limited. Two caveats to these conclusions are in order.

First, this analysis – based on hourly flow and capacity measures – uses a time unit that is likely to be too coarse to do justice to the potential benefits of ETC. On the other hand, the analysis of bottlenecks and resulting queues sug-gests that temporary disturbances to the roadway capacity can generate reduc-tions in the level of service that last much longer than the average disruptive event. Consequently, a much finer time unit could lead to increased benefit estimates. Second, this static analytic model makes specific assumptions on existing peak-hour travel demand. It is clear though that the time distribution of travel can change and the total volume of traffic can also change, thus undermining the basis for these conclusions.

Effects of ETC on Regional Network Congestion

The effects on the overall pattern of travel demand, which can be attributed to the increased capacity of a set of links in a network, depend on substitution and income effects stimulated by the changes in the relative cost of alternative routes at various times of the day. However, travel demand (and its hourly variations) is influenced by numerous other factors. Therefore, the impact of ETC should be isolated from the consequences of other phenomena. In this respect, it is useful to distinguish between the potential effect of ETC on the total level of travel demand and motorists' route choice decision.

The effects of ETC on travel demand include changes in the number of trips holding constant the spatial distribution of residential and business centers, as well as changes associated with future residential and business location decisions. Although these decisions are believed to be influenced by the existence and user costs of the local transportation network, quantifying this relationship with sufficient precision has proved to be very difficult – as indicated by the earlier discussion of traffic growth in the Dulles Corridor. Consequently, the accuracy of long-term forecasts of travel demand may be quite poor.

The effect of ETC on route choice decisions reflects the location of residences and business, but is also prone to be influenced by transient contextual factors that change the relative cost of alternative routes (for example, temporary alterations of capacity due to road work or accidents, and weather conditions). The installation of ETC alters the structural features of the network (by increasing the capacity of particular links) and thus can be expected to influence the basic parameters of the route choice decision. Insofar as ETC will be used for traffic management purposes (by adopting congestion pricing), it could also influence the short-term network conditions – and thus route choice decisions. This use of the ETC is not currently planned in either of the two cases analyzed here.

Rather than attempting to formulate quantitative estimates of the effects of ETC on patterns of travel demand and network traffic conditions, it is worthwhile considering what lessons can be drawn from the recent experience in the Dulles Corridor. Drawing an analogy between ETC implementation and conventional increases in highway capacity, the recent experience with the latter can be used to evaluate whether the former will restore the balance between the transportation needs of the Dulles Corridor and network functionality.

The historical trends in regional traffic growth indicate that the last 15 years have seen a tremendous increase in travel demand across the Dulles Corridor. This resulted from the growth of resident population in parts of Loudoun County (most notably the area around Leesburg and the settlements in Herndon and Sterling) and the growth in employment centers located in the western section of the Dulles Corridor (Reston Business District and Route 28). Consequently, the pattern of travel demand is heterogeneous, with commuter traffic moving to and from the Washington Metropolitan Area representing only one important factor behind a.m. and p.m. rush-hour congestion.

Despite the sustained growth of daily traffic on the DTR (see Tables 3.5 – 3.7) and its expansion to six lanes in 1992, travel demand along alternative routes in the west–east direction – namely, Rt. 7 and Rt. 50 – did not decrease. Traffic flow trends on Rt. 7 display a similar pattern to that of the DTR (Table 3.5), with higher growth on the Loudoun County part of Rt. 7 and declining growth on the Fairfax County segments. The DTR has thus attracted a share of the traffic originating from Loudoun County and directed to destinations in

Fairfax County and beyond. However, the overall growth in the number of trips along the Dulles Corridor dominated the potential congestion-reducing effect of the DTR.[17] This interpretation is supported by the analysis of traffic growth on Rt. 28 (running in the north–south direction between Rt. 7 and Manassas). Traffic grew at about 30 percent per year from 1986 to 1991 in the section north of the Dulles Airport and the DTR.

This interpretation is supported by the analysis of the growth of traffic volumes on Rt. 28 (running in the north–south direction between Rt. 7 and Manassas). The volume of traffic in the section north of the Dulles Airport and the DTR has grown at about 30 percent per year between 1986 and 1991. Conversely, there was no significant diversion of traffic from Rt. 50 to the DTR. In fact, while the link between Rt. 50 and the DTR experienced traffic growth on the order of 13 percent per year in the same time period, the traffic on Rt. 50 has grown more rapidly east of Rt. 28. It appears more plausible then that the traffic flow there is between Loudoun County and the southern parts of Fairfax County, Arlington County and further south.

These trends show that the growth of travel demand has proceeded well ahead of network capacity. Traffic intensity measures (for example, DSV) have consistently approached or exceeded desirable levels on the major arterials. The increased capacity provided by the opening and successive widening of the DTR, as well as other improvements on Routes 7, 28, and 50, has not reduced network congestion through the reallocation of traffic flows across the links. Current forecasts assume that the growth of travel demand in the region will continue at a lower pace in Fairfax County but at a higher one in Loudoun County. It is possible then that the congestion-reducing benefits from the opening of the DG and the two ETC installations may again be countered by the increase in total traffic in the network.

In summary, the notional advantages in the technical performance of ETC relative to other modes of toll collection may not translate completely into actual benefits from the standpoint of congestion and delay on transportation networks. It should be pointed out, however, that the relationship between the pattern of ETC implementation and the overall efficacy of the transportation network may be characterized by significant non-linearities. Accordingly, the effects of ETC on network capacity may display increasing returns to their spatial diffusion.[18] Thus, it is noteworthy that even the limited benefits of localized ETC installations can provide highway operators with sufficient incentives for their piecemeal adoption.

Interorganizational Coordination and Technical System Compatibility

Coordination of the technical choices made by the highway operators – VDOT and TRIP II – has been necessary to achieve the benefits of compatibility

Table 3.5   Annual average daily traffic on Route 7

	1981	2005 for.	1986–7	2010 for.	1991	1993
Rt7bW to Rt15W	8 480	17 000	16 940	25 684	21 000	22 000
Rt15W to Rt15E	8 480	17 000	26 000	47 000	24 000	26 000
Rt15E to Rt641	13 970	22 440	21 640	44 607	43 000	45 000
Rt641 to Rt28	10 190	22 440	21 640	44 607	41 000	48 000
Rt28 to Frfx Cl	11 300	40 000	27 265	50 336	54 000	56 000
Lou Cl to Rt717	29 940	45 000	44 590	86 613	52 000	52 000
Rt717 to Rt123	31 770	51 000	44 590	86 613	49 000	53 000
Rt123 to I495	56 440	74 000	61 880	105 907	60 000	57 000
I495 to F.C.	25 630	42 000	40 530	79 531	34 000	29 000

Source: VDOT

Table 3.6   Annual average daily traffic on Route 28

	1981	2005	1986–7	2010	1991	1993
Rt29 to I-66	18 050	35 000	25 545	63 924	50 000	60 000
I-66 to Rt50	6 570	28 000	11 340	30 949	24 000	26 000
Rt50 to Rt267	9 760	40 000	16 245	60 000	34 000	
		17 820				
Rt267 to Rt625	9 760	15 310	17 830	38 000	47 000	51 000
Rt625 to Rt7	9 760	15 310	1 7830	38 000	40 000	45 000

Source: VDOT

Table 3.7   Annual average daily traffic on Route 50

	1981	2005	1986–7	2010	1991	1993
Rt606 to Rt28	9 720	30 000	13 590	32 357	14 000	15 000
Rt28 to Rt645	13 210	38 500	30 560	80 385	44 000	43 000
Rt645 to Rt608	13 210	38 000	40 980	115 675	49 000	52 000
Rt608 to I-66	13 210	38 000	40 980	115 675	53 000	55 000
I-66 to WCL F	37 410	55 000	50 140	126 197	58 000	48 000
ECL F to I-495	32 840	50 000	34 755	54 259	44 000	47 000

Source: Virginia Department of Transportation (1989)

between the ETC systems implemented on the DG and the DTR. This commitment created constraints for the agencies involved at the same time that the novelty of the technical system posed challenges of its own to their procurement procedures. While technical compatibility between ETC systems was ultimately ensured, several problems and costly mishaps occurred along the way.

The introduction of ETC technology on the DTR was first considered in 1988 and a first round of proposals was let in 1990. However, only two companies (Amtech and Automatic Toll Systems) submitted their bids to VDOT, so the procurement process was suspended. In the summer of 1991 a second request for proposals (RFP) was issued to which four companies responded – Cubic Western, Science Applications International, Kiewit Network Technologies, and Westinghouse. Three of the four bidders for the $12 million contract included automated vehicle identification (AVI) equipment by Vapor Canada (a unit of Mark IV Industries) in their bids.[19] The outcome of the proposal appeared so obvious that the Toll Road Corporation of Virginia (TRCV), the private partnership that started the development of the DG, announced their selection of Vapor's AVI equipment even before the award of the VDOT contract.[20]

Having failed to reach an agreement with Cubic, VDOT awarded the contract to Kiewit Network Technologies (now MFS Network Technologies), whose proposal included AVI equipment from Texas Instruments. After Cubic filed a lawsuit against VDOT because the procured system did not meet the specification, VDOT invoked other irregularities to scrap the procurement altogether (*Inside IVHS*, June 21, 1993), forcing TRIP II, the new partnership behind the DG development, to follow suit by canceling its own agreement with Vapor.[21]

VDOT's third RFP, launched in the fall of 1993, was answered by proposals from six companies (Cubic, Motorola Government and Systems Technology Group, Amtech, MFS, Science Applications International, and Maris Equipment). VDOT awarded the $11.8 million contract to Cubic Toll Systems, a system integrator who had offered to install the Premid ETC system from a Swedish firm, Combitech Traffic Systems. However, MFS Network Technologies challenged the procurement decision in a court, claiming the procedures were flawed and unfair (*Inside IVHS*, September 12, 1994). Before the lawsuit went to trial, TRIP II awarded a $8.6 million contract for the deployment of the ETC system to Syntonic, a subsidiary of Science Applications International Corporation (*Inside IVHS*, July 18, 1994).

Before the trial ended, a settlement was brokered between VDOT, Cubic and MFS. VDOT agreed to test MFS's AVI equipment (developed jointly with Texas Instrument) and, upon verification that it met the procured specifications, to request that Cubic (a system integrator) install MFS's equipment rather than Combitech's Premid, as from the original proposal. However, MFS's equipment failed the test and Cubic could proceed with its own proposal.

While favoring Combitech's Premid system, Cubic had actually offered an alternative one in its proposal, developed by Mark IV, that meanwhile was selected by Syntonic, the prime contractor for the DG's ETC implementation. The two systems were not compatible.[22] At last, VDOT decided to adopt the same system picked by the contractor of the DG, namely Mark IV. Combitech delivered the equipment procured by VDOT. In September 1995, VDOT released Cubic Technology from its obligations after paying $5 million for the work done, and awarded a $9 million contract to Syntonic for the completion of the ETC installation (Cooper, 1995).

CONCLUSIONS

The evaluation of the ETC deployment projects at the DTR and the DG according to the checklist methodology has made it possible to provide an account of the expected outcomes along a variety of dimensions. Given the nature of the technology the emphasis has naturally been placed on the effects it will have on traffic patterns in the region. The occurrence of higher-order effects on residential and business development of the region hinges on the contribution that the new technology will make to the performance of the local transportation network *vis-à-vis* current and future travel demand.

The most important lesson that can be drawn from this case study concerns the need to look into the details of the ETC implementation sites in order to clarify the extent to which time- and site-specific factors influence the economic value of the expansion of toll lane capacity implicitly brought about by ETC. This evaluation, carried out by implementing a heuristic bottleneck analysis, indicates that the existing location of toll plazas and the capacity of adjacent highway segments may significantly hamper the ability of ETC to enhance traffic flows. This will be the case whenever the toll plaza does not constitute the most severe bottleneck in an appropriately defined portion of the network. It has also been argued that the pervasive presence of external effects (positive and negative) in a transportation network interferes with the magnitude of the contribution that localized expansions of capacity can make to network performance. Based on the foregoing analysis, it appears that the installation of ETC on the DTR and the DG will have small positive effects toward reducing congestion during peak hours, and larger positive effects toward reducing travel time during off-peak hours. Lower travel demand during off-peak hours implies lesser congestion and the almost complete absence of external effects amongst individual motorists. The benefits of ETC adoption are in this case completely internalized by the individual users.

In this case study we have analyzed the process by which the implementation of the technology has occurred. Given the geographic proximity of the two highways, the necessity for appropriate governance structures to coordinate the choice of the ETC system, its procurement and deployment was particularly strong. The combination of public and private sponsors to these projects provided a peculiar instance of the inter-agency coordination that is expected to favor the effective deployment of ITS technologies. In this case, the record has been pretty disappointing. The development of the ETC systems lagged behind schedule, and the coordination between the two operators to guarantee the inter-operability of the AVI tags only emerged at the end of a series of controversies. Although other factors – not necessarily linked to the relationship between the two agencies – have influenced the process of selection, procurement and development of the ETC systems, the governance structure realized here has not fared very well, particularly in comparison to other instances (for example, the IAG in the New York area).

APPENDIX: ELECTRONIC TOLL COLLECTION CHECKLIST

Category	Indicator	Effect	Comments
Congestion and delay	Peak-hour VMT	Increase	The principal benefits of ETC can be expected in the reduction of congestion and delays by reducing queues at toll plazas and smoothing traffic flows around peak-hours. The benefits will decrease when congestion forms downstream from ETC installations.
	Vehicle-hours of delay	Decrease	
	Peak/non-peak travel times	Decrease	
	Average speed	Increase	
	Average length of queue	Decrease	
Capacity and mobility	Ease of personal mobility	Increase	Shorter trip duration.
	System accessibility	—	—
	Network or connectivity implications	Increase	May reduce congestion of alternative non-tolled network links.
	Daily VMT	Increase	Redistribution of traffic flows from other routes.
Safety and accidents	Deaths	No effect	Speed limits and dedicated lanes for ETC should prevent increased risk of accidents.
	Injuries	No effect	
	Property damage	No effect	
Cost relative to other options	Construction	Decrease	ETC creates additional toll lane capacity at lower cost than adding new lanes at toll plazas.
	Operations	Decrease	Based on experience of existing ETC installations.
	Maintenance	Decrease	

Category	Indicator	Effect	Comments
Financial and fiscal outcomes	Budgetary sufficiency	Positive	Reduced operating costs should improve margins, generating investment funds for VDOT.
	Inter-governmental transfers	None	DTR project financed by VDOT's internal funds; DG project is private.
	Private profits	Positive	DG investors expect to profit before expiration of their license.
Environmental outcomes	Total emissions	Decrease	Reduced idle time at toll plazas and congestion elsewhere in the network.
	Ground level ozone	Decrease	
	Energy consumption	Decrease	Reduced idle time and fuel efficiency from improved traffic flows at toll plazas.
Economic outcomes	Employment	Indirect positive effect	No direct effect from installation work, positive indirect effect through acceleration of economic development of outlying areas.
	Per capita income	No effect	
	Land values	Positive effect	Enhanced mobility will increase attractiveness of region for residential and commercial development.
	Regional competitiveness	No effect	
Equity outcomes	Affordability	Negative effect	DG toll is high and increased recently. No change in tolls charged for DTR.

Category	Indicator	Effect	Comments
	Regional access disparity	Increase	DG tolls may lead to segregation of commuting patterns by income level.
	Handicap access	—	
Institutional outcomes	Flexibility and adaptability	Positive effect	ETC creates conditions for adoption of congestion pricing.
	Regional significance	Yes	Regional ITS strategy for dealing with congestion problems.
	Community involvement and preferences	Yes	Both projects received input from resident and business communities.
	Public–private partners	Yes	Cooperation between DG operator and VDOT necessary for inter-operability.

NOTES

1. These two projects – the first of this kind in the Commonwealth of Virginia and the Washington Metropolitan Area – were undertaken without financial support from the Federal Government, support that has been available in other regions of the US to stage operational tests for ETC as well as other ITS technologies. Installation on the DTR has been sponsored by the Virginia Department of Transportation using agency's general investment funds – the DG is entirely financed by private funds. Together these projects provide an excellent instance of the variety of institutional arrangements leading to the deployment of ITS technologies proposed in the National ITS Program Plan (Euler and Robertson, 1995). They also aim at realizing the goals that the latter identified in the "more efficient use of our infrastructure and energy resources, and significant improvements in safety, mobility, accessibility, and productivity" (p. 1).

2. Non-agricultural employment rose by 75 percent over the 1980s in the Virginia portion of the DC Metropolitan Statistical Area. The major sources of growth have been increased traffic at the Dulles Airport, outsourcing of federal defense procurement, and growth in the technology sector. The Airport's workforce increased at 11.3 percent per year over the 1980s and it was estimated to continue growing albeit at a lesser rate of 7.3 percent during the 1990s. Defense prime contracts awarded to Virginia-based firms rose from $3.3 billion in 1980 to more than $10 billion in 1988, with Fairfax County taking in a large share of the total. About 70 percent of the contract awards went to firms in the service sector.

3. At the time of development the DTR was designed to accommodate 46 000 vehicles per day, although the Virginia Statewide Highway Plan (VDOT, 1989) considered the acceptable daily service volume for the DTR to be 54 200 vehicles per day (81 300 in the segment between Route 7 and the Beltway).

4. According to the Virginia Statewide Highway Plan, the ratios of average daily travel to the DSV for different segments of the highway ranged between 0.25 and 1.01 (VDOT, 1989).

5. The *Highway Capacity Manual* (Transportation Research Board, 1994) defines the levels of service as an indicator of road traffic conditions based on vehicle density. The levels range from A (low density of vehicles) to F (very high density and congested traffic).

6. The DTR was in fact built on right of way belonging to the Washington Metropolitan Area Transit Authority (WMATA) which acquired it in the early 1960s in order to build the Dulles Airport Access Road. The Virginia Department of Transportation (VDOT) leased the right of way from WMATA under a variety of concessions, including clauses according to which a fraction of the toll revenues had to go toward the development of a light-rail link to be laid out in the median of the highway.

7. The motivations for the legislation were somewhat broader than the specific issues related to the traffic conditions in the area where the DG was to be built. Rather, the DG was seen as the first initiative demonstrating the economic and political viability of private investment in transportation infrastructure.

8. Loudoun County's local authorities envisaged the project as an important stimulus to the economic development of the region. The areas adjacent to the DG are characterized by scattered residential communities (especially in the western section) amidst farmland, forested and open land, in spite of the rezoning for development in the spaces contiguous to the Dulles Airport. Local planning authorities continue to consider the growth of the Dulles International Airport as a primary stimulus to business and industrial development with the DG serving as the backbone of the local transportation infrastructure (Loudoun County Board of Supervisors, 1995, p. v).

9. Rephann and Isserman (1994) have used quasi-experimental matching techniques to test the relationship between economic growth and investment in interstate highways at the county level. Their findings are very interesting and suggest that the main beneficiaries of the interstate system in terms of economic growth are the areas in close proximity to large cities or with some degree of prior urbanization.

10. This experience could be compared with that of the Inter-Agency Group (IAG) that was formed among highway and bridge operators of the New York area to jointly decide upon

the technical standards of the ETC system to be deployed (see Gifford *et al.*, 1996).
11. Bottlenecks that locally provide the smallest capacity are binding. The opposite concept is that of a hidden bottleneck.
12. A common abstraction in the analysis of motorists' route choice in simple networks is to consider an equilibrium allocation of travel as the distribution of traffic across links in the network such that a motorist should be indifferent at the margin between alternative routes. However, in the more realistic setting of complex networks (with multiple origin–destination pairs) an equilibrium cannot be defined in this way, since for each O–D pair it is possible that one route will strictly dominate all the others available (Arnott *et al.*, 1993).
13. Notice how regulating this rate is the objective of ramp metering.
14. During peak hours traffic counts on segments of the DTR between SR675 and SR676 average around 6420 (a.m.) and 6600 (p.m.) vehicles.
15. The measure is the traffic volume at the entry or exit point selected (Sully Road for the 1994 traffic counts) as a proportion of traffic across the mainline toll plaza.
16. The issue of the time unit for capacity measurement (discussed in the next paragraph) is crucial here. Based on the standard measurement as vehicles per hour per lane, the travel demand is likely to be always below the capacity of the toll plazas under manual collection. Strictly, the toll lanes would not constitute a bottleneck at all.
17. A report of the Fairfax County Office of Transportation sets at 110 percent the 10-year (1981–91) increase of traffic flows across the Loudoun County borders, an estimate which is consistent with the MWCOG estimate of the same period growth of home-based work trips for Loudoun County (86 percent).
18. Consider the case of two identical bottlenecks on a highway segment. Investment in additional highway capacity at either of the two bottlenecks will not reduce total travel time, but investment at both will do so.
19. The second proposal contained technical specifications for the AVI equipment allegedly modeled after those of the HELP project (*Inside IVHS*, January 6, 1992) and perhaps were wired involuntarily toward Vapor's equipment (Fehr, 1990).
20. The agreement negotiated between TRCV and VDOT to grant the former with the permission to start building the DTR extension to Leesburg included provisions to the effect that the two parts of the highway should be seamlessly linked together from the perspective of the users. This was understood to mean that the selection of the AVI equipment by the TRCV had to guarantee inter-operability across the DTR and its extension.
21. TRIP II was not bound to procure the ETC system or its components from the same vendor chosen by VDOT. At the end of 1993, however, there were no compatible systems available on the market: the agreement translated *de facto* into a commitment to purchase the equipment from the same vendor. The first announcement by two companies concerning the development of compatible technologies occurred in February 1994. The two companies were Hughes Transportation Management Systems and Mark IV Industries.
22. At the most basic level, the difference lay in the frequency of the vehicle-to-roadside communication which was in the 2.45 GHz band for the Combitech system and in the 902–928 MHZ band for the Mark IV system. Thus, the latter system operates in the frequency band that has been assigned to ETC systems by the Federal Communications Commission (FCC). Combitech's system on the other hand reflects the dominant choice of frequency band for European providers. In addition to questions concerning the legal implications of installing an ETC system operating on a frequency outside the FCC designated band, the main problem was the lack of inter-operability across the two highways (*Inside IVHS*, March 13, 1995).

REFERENCES

Arnott, R., De Palma, A. and Lindsey, R. (1993), "Properties of dynamic traffic equilibrium involving bottlenecks, including a paradox and metering", *Transportation Science* **27**(2), 148–60.

Arthur, B. (1990), "Positive feedbacks in the economy", *Scientific American*, February, 92–9.

Braess, D. (1968), "Uber ein Paradoxen der Verkehrsplanning", *Unternehmensforschung* **12**, 258–68.

Cooper, A. (1995), "Dulles toll road project delayed by judge; awarding of contract was challenged", *Richmond Times Dispatch*, December 8, B-3.

Euler, Gary W. and Robertson, H. Douglas (eds) (1995), *National ITS Program Plan: Synopsis*, Washington, DC: ITS America.

Fehr, S.C. (1990), "Va. accused of favoritism in toll-collecting project", *Washington Post*, Metro Section, July 3, B5.

Forecasting in the Washington Region (Summer 1998), Publication no. 98817. http://www.mwcog.org.

Gifford, J.L., Yermack, L. and Owens, C. (1996), "EZ Pass: A case study of institutional and organizational issues in technology standards development", *Transportation Research Record* 1537 (November), 10–14.

Gómez-Ibáñez, José A. and Meyer, John R. (1993), *Going Private: The International Experience with Transport Privatization*, Washington, DC: Brookings Institution.

Inside IVHS, various issues.

JHK & Associates (1995), "Western Washington Bypass Major Investment Study – Truck Data Memorandum", Alexandria, VA.

Loudoun County Board of Supervisors (1995), *Toll Road Plan*.

Metropolitan Washington Council of Governments (1998), "Growth Trends to 2020: Cooperative Forecasting in the Washington Region", Publication no. 98817. http://www.mmcog.org.

Mohring, H. (1999), "Congestion", in Jose A. Gómez-Ibáñez, William B. Tye and Clifford Winston (eds), *Essays in Transportation Economics and Policy*, Washington, DC: Brookings Institution Press, pp. 181–221.

Pietrzyk, Michael C. and Mierzejewski, Edward A. (1993), *Electronic Toll and Traffic Management (ETTM) Systems*, NCHRP Synthesis 194, Transportation Research Board, National Research Council, Washington, DC: National Academy Press.

Pigou, Arthur C. (1920), *The Economics of Welfare*, London: Macmillan.

Plane, D.A. (1995), "Urban transportation: policy alternatives", in Susan Hanson (ed.), *The Geography of Urban Transportation*, New York: Guilford Press, pp. 435–69.

Rephann, T. and Isserman, A. (1994), "New highways as economic development tools: an evaluation using quasi-experimental matching methods", *Regional Science and Urban Economics* **24**(6), 723–51.

Stough, R., Maggio, M. and Jin, D. (1994), *Methodological and Technical Challenges in Regional Evaluation of ITS Induced and Direct Effects*, Report to US Federal Highway Administration, October 1994.

Transportation Research Board (1994), *Highway Capacity Manual*, Washington, DC: National Research Council.

Vickrey, W. (1969), "Congestion theory and transport investment", *American Economic Review* **59**(3), 251–60.

Virginia Department of Transportation (1989), *Statewide Highway Plan*, Transportation Planning Division.

4. The Variable Message Sign System of Northern Virginia

Brien Benson

CONTEXT AND ITS APPLICATION

Electronic variable message signs (VMSs) are a means of communicating traffic instructions and traffic and road condition information to motorists en route. VMS are currently deployed in about a dozen of America's largest metropolitan areas and in many overseas cities.[1]

The most common use of VMSs in the United States is to provide current travel conditions, such as traffic congestion, weather-related road conditions, and notification of construction ahead. VMSs may also give traffic directions, such as controlling access to high-occupancy vehicle (HOV) lanes or announcing detour instructions. VMSs are also used for safety messages, notice of future road construction, and notice of public events.

This study considers 100 VMSs in Northern Virginia. These are under the control of the Virginia Department of Transportation (VDOT), acting through its Arlington, Virginia, Transportation Management System (TMS). Some 60 of these signs are fixed VMSs, placed overhead on the three interstate highways of the region – I-66, I-95/395, and the Capital Beltway. Some 40 portable VMSs are available for deployment along either interstate or arterial highways. Another 60 fixed VMSs are scheduled for deployment along interstates in the next few years.[2]

The VMSs provide traffic and road condition reports, instructions for using HOV lanes, instructions for using alternate routes in the case of major traffic tie-ups, safety messages, and time-and-date displays. The primary source of VMS traffic information is loop detectors embedded in highways, but some 40 closed-circuit television cameras throughout the region are available to verify information generated by loop detectors, and provide additional information, such as weather conditions and the causes of congestion.

The system has a number of problems which limit its overall effectiveness. First, the equipment is somewhat dated. The central computer is 25 years old. Coordination with other sources of traffic information is not yet automated, although steps in this direction are being taken. The closed-circuit television cameras are sometimes hard to control. And the loop detectors were installed in already-existing pavement, a procedure which tends to exacerbate maintenance problems.

Second, staff are not as well trained as they might be (as described by Miller *et al.*, 1995, p. 2).[3] Knowledge of the region's transportation network is limited. More fundamentally, with the exception of the director of the TMS, the overall level of professionalism is not high.

Third, the region's road network offers a limited number of useful alternate routes in case of freeway congestion.[4] Thus, even if VMSs provide accurate and timely information about freeway traffic conditions, the options available to motorists are limited.

Finally, when the VMSs were originally deployed in the mid-1980s, their primary purpose was to control access to HOV lanes. As a result, the VMSs were sometimes placed in locations that are illogical from the perspective of motorist information needs. For example, some VMSs are located immediately after a decision node – that is, after the last point at which a motorist could choose to exit the freeway, or not to enter the freeway.

EVALUATION ISSUES

There are a number of difficulties inherent in carrying out an objective, quantitative evaluation of any VMS system. First, it is difficult to measure the impact of VMSs on driver behavior.[5] For example, if a VMS announces "Congestion Ahead", it is virtually impossible to ascertain what proportion of cars exiting the freeway during the intervening distance did so as a result of the VMS message.

Second, even if we could identify which motorists respond to VMS messages, we cannot necessarily determine how much time they saved from their change in routes.[6] To do this would require comparing the drive times along the alternate and the original route in every case, and then summing the differences. Other benefits, such as lower incidence of accidents, are even harder to quantify.

Third, even if travel time savings or accident prevention could be measured, different individuals are likely to place a different dollar value on these benefits. Finding an agreed-upon methodology for such dollar valuations has proven very difficult. Studies have emphasized quite different criteria for evaluation, including the traveler's income (American Association of State Highway and Transportation Officials, 1978), the purpose of the trip (Ruhl, 1993), the length of the trip (Tretvik, 1992), the presence of other riders in the vehicle, and even the family size of the traveler. The plethora of studies has not led to any consensus.

A final difficultly in evaluating VMSs is that some of their functions might be performed nearly as well by a traditional sign, for example, listing the times when HOV lane controls are in force. This complicates evaluating "how well" the VMS does its job in such situations.

All these difficulties in evaluating VMS impacts on traffic flow are magnified enormously when trying to assess second- and third-order effects of VMSs, such as impact on land use and on economic development.[7] In such cases the collection of meaningful data is extraordinarily difficult.

In view of these difficulties in devising objective measures of the impact of VMSs on travel, it seems that a more subjective methodology – that of public opinion analysis – may be the most promising approach. While the "stated preferences" of opinion surveys have obvious limitations compared to "revealed preferences" of objective studies, in cases like this they may be the only reasonable option available. The results of the application of an alternative, comprehensive evaluation methodology – the "benefit assessment checklist methodology" – is presented in the Appendix to this chapter.

EVALUATION OF THE ITS APPLICATION: VMSs IN NORTHERN VIRGINIA

The public opinion survey reported in this section was conducted during the summer of 1995 as part of a project to assist the Virginia Department of Transportation (VDOT) in managing its Northern Virginia VMS systems. There were 520 interviews, and respondents' residences ranged from Fredericksburg, Virginia, to Rockville, Maryland.[8] Seven focus groups were held first: to help in formulating the survey questions, and to provide a depth and qualitative dimension to public opinion that could not be achieved in the survey.

The survey was more concerned with possible improvements that could be made to VMSs than with an evaluation of the existing system, and so was limited in the number of questions it could ask about current services. Nonetheless, the survey did provide some useful evaluation information. In addition, it provided a way to test the public opinion methodology of evaluation. In the next section of the chapter, the results of the analysis of the opinion survey are presented.

Influence of VMSs on Driver Behavior

When asked how often they are influenced by VMSs, half of the survey respondents (254) said "often"; two-fifths (195), "occasionally"; and the rest (67), "not at all".[9] Stated differently, half the respondents rely regularly on

VMSs, and half do not. This 50–50 split in opinion in the survey respondents is consistent with the views of the focus groups, where opinion on VMSs was roughly evenly divided between positive and negative.

Sixty percent of respondents said they would be "very likely" to use VMS-recommended alternate routes posted in case of severe congestion ahead, tending to confirm that roughly half of all motorists consider VMSs an important source of information.

Accuracy of Traffic Information

Only one-third of survey respondents said they had experienced inaccurate or out-of-date information on VMSs. This seems like a low percentage, because it is a fact that VMSs in the region are sometimes inaccurate. Indeed, in the focus groups there was widespread discussion about inaccuracies on VMSs. One possible explanation for this anomaly is that motorists relying on VMSs only "occasionally" or "not at all" did not notice VMS inaccuracies.

Value of Safety Messages

Two-thirds of survey respondents think it is a "good idea" to post safety messages, such as "drive to survive" and "lights on in bad weather". Support was considerably stronger for specific messages like "signal before changing lanes" and "lights on in bad weather" than for general messages like "drive to survive" and "tailgating is deadly".

This high level of approval for safety messages was inconsistent with the focus group responses, whose participants generally preferred that VMSs be limited to current traffic information and traffic instructions It is possible that a different phrasing of the survey question – for example, "Do you think VMSs should display only current traffic and road conditions, or do you think safety messages should also be displayed?" – would have yielded less support for safety messages.

Suggested Improvements

The study found strong support for a number of suggested improvements to the Northern Virginia VMS system:

- Motorists strongly supported the idea of posting the exact location of an accident ahead, so they would best know how to drive around it.
- Motorists liked the idea of an "anti-rubbernecking" message. This would read "Accident Ahead: All Lanes Clear: Please Maintain Speed" and would be posted when the after-effects of an accident can be seen along-

side the road, but all travel lanes have been cleared.

- Most motorists favored the idea of "Time-tagging" traffic information, that is, displaying the time when a traffic report was first posted on a VMS. The idea behind "time-tagging" is to help motorists decide whether they think a VMS traffic report is sufficiently timely to be accurate.
- Several participants in the focus groups suggested that VMSs be placed at better decision nodes, and several participants felt there should be more VMSs. Other suggestions included putting portable VMSs at key accesses to important surface streets, and dispatching portable VMSs to crisis locations, such as major accidents.

These findings suggest that many motorists think the VMS system is essentially worthwhile but needs to be managed better.

Influence of Demographics

In general, no significant relationship between motorist attitudes and demographic variables was found. There were some exceptions to this statement, particularly with regard to education. A small negative correlation between educational attainment and support of VMS safety message and a negative correlation of -0.179 between education and preference for display of television scenes of traffic on VMSs were found. A similar negative correlation was found between education and support for displaying maps on VMSs.

LESSONS LEARNED AND TRANSFERABILITY OF FINDINGS

What, then, may be concluded about the benefits of VMSs, based on the Northern Virginia study of motorist attitudes? Probably most important is that roughly half of the region's motorists rely regularly on VMSs. In the face of all the limitations of Northern Virginia VMSs, this seems like a rather impressive level of support. At the same time, motorists appear to think VMSs can be made more useful. The survey respondents called for better accuracy and timeliness, more useful safety messages, improved methods of presenting information, and additional types of information. Motorists appear to be saying that the VMS system has not achieved its potential. Significantly, the benefits of VMSs are seen similarly by all groups of motorists, whatever their place of residence, income, age, gender, or education. In other words, there do not seem to be any major equity or distributional problems plaguing VMSs.

Is it possible to relate these conclusions to the primary goals of the IVHS Act of 1991: improved capacity, efficiency and safety of the federal-aid highway system, and progress towards cleaner air? Based on the motorist survey, it is difficult to draw conclusions specific to the four primary goals of the IVHS Act. However, it seems clear that the methodology that has been developed could be adapted and a new survey conducted that would be able to draw such specific conclusions. What, finally, may be said about the transferability of information learned from this study? To the degree that clear sets of attitudes towards VMSs have been identified, it seems reasonable to assume that such attitudes in other US metropolitan areas would not differ dramatically from what was found in Northern Virginia. American culture, particularly American automobile culture, is fairly homogeneous throughout the country. Furthermore, demographic characteristics do not seem to affect attitudes towards VMSs, so, even if different regions have different demographic make-ups, their attitudes towards VMSs would not tend to be affected.

Of course, different regions have different VMS systems, to which people will react differently. And different regions have different road networks, which affects the usefulness of VMS-directed re-routing of traffic. Such differences could lead to important differences in public attitudes towards VMSs. But, if such technical differences are kept in mind, it seems reasonable to assert that the findings of this Northern Virginia study are at least generally transferable to other metropolitan regions in the US.

APPENDIX: NORTHERN VIRGINIA VARIABLE MESSAGE SIGNS CHECKLIST

Category	Indicator	Effect	Comment
Congestion and delay	Peak-hour VMT	Increase	The principal benefits of VMSs lie in the category "Congestion and delay", by smoothing out traffic flow through re-routing traffic around congested areas. Perhaps the most practical way to assess these benefits is through motorist surveys.
	Vehicle-hours of delay	Decrease	See above.
	Peak/non-peak travel times	Decrease	See above.
	Average speed (e.g., PMT/P-Hr.)	Increase	See above.
	Average length of queue	Decrease	See above.
Capacity and Mobility	Ease of personal mobility	Increase	—
	System accessibility	Increase	VMSs on approach routes to freeways could rationalize motorist use of freeways.
	Network or connectivity implications	Increase	Rerouting of traffic will tend to increase network utilization.

Category	Indicator	Effect	Comment
	Daily VMT	Increase	Slight increase due to increased throughput on freeways.
Safety and accidents	Deaths	Decrease	All safety indicators should improve with smoother traffic flow promoted by VMSs.
	Injuries	Decrease	See above.
	Property damage	Decrease	See above.
Cost relative to other options	Construction	Variable	No other options are strictly comparable. While HAR and commercial radio broadcastscan provide traffic information, only VMSs can be used for legally enforceable traffic instructions, and can be assumed to have been read by all motorists passing a given location. It should be noted, however, that at present, private sector alternatives provide some level of traffic information at no cost to the taxpayer, and additional private sector options are being developed.
	Operations	Variable	See above
	Maintenance	Variable	See above
Financial and fiscal outcomes	Revenue enhancement	No effect	—

Category	Indicator	Effect	Comment
	Budgetary sufficiency	Moderate budgetary demands	Upgrade of system and of staffing, and improved maintenance, could improve performance substantially. But such improvements face severe budgetary constraints.
	Inter-governmental transfers	None	—
	Private profits	None	—
Environmental outcomes	Total emissions	Probably beneficial	Analysts do not fully agree on models for what causes emissions, but the smoother traffic flow from VMSs is probably beneficial for air quality.
	Ground level ozone	Probably beneficial	See above.
	Energy consumption	Probably beneficial	Increased throughput, and therefore increased VMT, permitted by VMSs would probably be more than offset by smoothing out of driving, inasmuch as start-and-stop driving is highly energy inefficient.
	Hazardous materials impact	Slightly beneficial	Any procedure that smooths out traffic flow diminishes chances for accidents, including hazmat accidents.
	Soil erosion	No effect	—
	Wetlands and habitat	No effect	—

Category	Indicator	Effect	Comment
Economic outcomes	Gross regional product	Increase	All indicators in this category should slow a slight increase, reflecting the more rapid traveling times and more pleasant driving experience due to VMSs. Efforts have been made to put a dollar value on travel time savings.
	Employment	Increase	See above.
	Per capita income	Increase	See above.
	Land values	Increase	See above. (There might be a slight decline in land values where freeway traffic is routed through local commercial or residential sites.)
	Regional competitiveness	Increase	See above.
Equity outcomes	Affordability	No effect	—
	Regional access disparity	Variable, but minor	Those using freeways most heavily will benefit most; it is not clear whether this would be residents of "inner" counties, residents of "outer" counties, or through traffic.
	Handicap access	No effect	—

Category	Indicator	Effect	Comment
Institutional outcomes	Flexibility, adapts	No effect	—
	Regional significance	Minor	Slightly increases authority of the state department of transportation, which manages the VMS system, vis-à-vis local authorities. However, the system must be managed with sensitivity to local needs.
	Synergy	No effect	—
	Inter-governmental friction/cooperation	No effect	VMSs are directly run by VDOT.
	Community involvement and preferences	Could be substantial	One proposed use of VMSs is to announce public meetings regarding transportation planning. If enacted, this use could significantly increase community involvement.
	Ownership changes	No	—
	Public–private partners	No	—
	Liability	Uncertain	Traffic direction by VMSs probably incurs no greater liability than existing traffic direction activities of VDOT.

Category	Indicator	Effect	Comment
	Privacy	Variable	Present methods of traffic surveillance have no negative privacy impacts. New techniques, for example, advanced CCTV, might have adverse impacts.
Spatial outcomes	Land use implications	Possible	If rerouting of traffic from freeways through neighborhoods became extensive, which is unlikely under current political conditions, new patterns of commercial, industrial and residential use might emerge.
	Land values	Possible impact	See above.
	Residential proximity issues	Possible impact	See above.
	Infrastructure design or geometry impact	None	—
Demonstration outcomes	Transferability	Good	Survey research found no correlation between demographic variables and attitudes towards VMSs. Together with the generally homogeneous automobile culture of the United States, this would suggest good transferability of our findings. However, differences in VMS systems and regional road networks must be borne in mind.
	Value as a model	Not good	The VMS system in question is rather dated.

Category	Indicator	Effect	Comment
	Suitability	Good	Spread of VMSs to other metropolitan areas is likely. Therefore, our research would seem quite suitable to the issues facing ITS technologies.

NOTES

1. A review of the technical aspects of VMSs can be found in *NCHRP Synthesis 237. Change-able Message Signs: A Synthesis of Highway Practice*, Washington, DC: Transportation Research Board, National Cooperative Highway Research Program. An overview of the importance of VMSs can be found in Dr Brien Benson's "Variable Message Signs: The ITS Technology to Watch", *ITS Quarterly* **IV**(2), 1996, Washington, DC: ITS America.

2. The Virginia Department of Transportation publishes regular updates of its business plan for the extension of ITS facilities. These are available from the Virginia Department of Transportation, ITS Program Group, Richmond, VA.

3. The following paragraph appears in Miller *et al.* (1995, p. 2): "Despite widespread use of VMSs, no practical guidelines for their operation exist, forcing VDOT employees and private contractors to make judgments, with little training or guidance, regarding how, when, and where to sue the VMSs. Some VMSs have been ineffective as a result of confusing word choices, lengthy messages, ambiguous messages, or even incorrect placement of the VMS."

4. Northern Virginia "fans out" from a comparatively short stretch along the Potomac River (the boundary with the District of Columbia) to what is roughly a semicircle with a radius of 40 miles and a circumference of 100 miles. Because of this configuration, the use of parallel arterials is uncommon. Furthermore, because the road system grew up in a rather haphazard fashion, several of the key arterials do not form straight spokes from hub to circumference. The best single description of the region's road system is published by the regional planning organization, the National Capital Region Transportation Planning Board, *1997 Update to the Financially Constrained Long-range Transportation Plan for the National Capital Region* (available from Metropolitan Washington Council of Governments, Washington, DC).

5. A general review of the difficulties of using observed behavior to analyze driver response to VMSs is provided in Richard H.M. Emmerink, Peter Nijkamp, Piet Rietveld and Jos N. Van Ommeren, "Variable message signs and radio traffic information: an integrated empirical analysis of drivers' route choice behavior", *Transportation Research, Part A, Policy and Practice* (1966), **30**(2): 135–53.

6. A list of studies of travel time evaluation is available in Brien Benson, "Market-factors affecting deployment of advanced traveler information systems: an annotated bibliography", unpublished manuscript, George Mason University, School of Public Policy, Fairfax, VA, 1993.

7. The matter of second- and third-level effects is treated in Chapter 2 of this volume.

8. The survey dialed 2685 phone numbers and reached a human 1010 times. Reasons for the 1675 failures to reach a human were as follows: business, 40 (2.4 percent), disconnected, 263 (15.7 percent), computer/fax, 39 (2.3 percent), need to dial a "1", 5 (0.3 percent), no answer, 597 (36 percent), answering machine, 619 (37 percent), busy, 100 (6 percent), other, 12 (0.7 percent). Of the 1010 times we reached a human, we were unable to reach the designated respondent 303 times. Thus, we had a potential respondent on the line 707 times. Of these 707 potential respondents, we completed interviews with 517 for an overall response rate of 73.13 percent. Phone numbers were selected by random digit dialing and random respondent selection within the household was accomplished.

9. Confidence intervals at the 95 percent level are no greater than +/–5 percent for questions answered by at least 500 people. Most of the responses discussed in this chapter fall into this category. Where responses are substantially fewer, the total is indicated.

REFERENCES

American Association of State Highway and Transportation Officials (1978), *A Manual on User Benefit Analysis of Highway and Bus-Transit Improvements, 1977*, Washington, DC.

Miller, John S., Smith, Brian L., Newman, Bruce R. and Demetsky, Michael J. (1995), *Final Report: Development of Manuals for the Effective Use of Variable Message Signs*, Charlottesville, VA: University of Virginia.

Ruhl, Aad (1993), "An economic theory of travel decisions", Transportation Research Board Annual Meeting preprint, Washington, DC.

Tretvik, Terje (1992), *Inferring Variations in Values of Time from Toll Route Diversion Behavior*, Transportation Research Record 1395, Washington, DC: Transportation Research Board.

5. Automatic Truck Rollover Warning System

Michelle Sager

INTRODUCTION AND PURPOSE OF THE ANALYSIS

The Automatic Truck Rollover Warning System (ATRWS) employs electronic sensing and communication technologies providing a visual warning to drivers on freeway exit (entrance) ramps who risk a possible turn over of their vehicle due to excess speed. The system uses roadside detectors and in-pavement weigh-in-motion sensors to ascertain the presence and speed of heavy trucks. If the truck exceeds the posted ramp speed (35 mph), lighted warning signs placed alongside traditional speed advisory signs are activated. This system was developed as a response to truck rollover accidents on highway exit ramps. Five out of every 100 fatal truck accidents in the US occur on off-ramps at interstate interchanges (Vallete *et al.*, 1981). Large trucks overturn under conditions in which a sharp radius (ramp) curve is preceded by stretches of road that encourage higher than safe approach speeds.

Truck rollover accidents usually result in: fatalities and injuries, vehicle and roadway damage, and traffic delays. Traffic delays produce second-order effects including decreased vehicle-miles traveled (VMT), and lost productivity. Moreover, trucks carrying hazardous cargo can cause environmental damage and greater losses when they are involved in truck rollover accidents. In short, the Automatic Truck Rollover Warning System is a response to a major highway safety issue.

The Automatic Truck Rollover Warning System deployed in the National Capital region is an example of an Intelligent Transportation System (ITS) deployment. The ATRWS represents an ITS project that has relied heavily on thorough but traditional cost-benefit analyses, while lacking evaluation of interaction effects and other factors that may affect its suitability as a model for deployment elsewhere.

This chapter seeks to expand the scope of analysis beyond the traditional cost-benefit assessment through the use of a multi-criteria approach to ITS assessment. The approach includes some traditional and essential elements (for example, a description of the project, its history, and the technology involved in its application) as well as a description of the traditional project evaluation methods and results. The case study then applies a multi-criteria checklist methodology to highlight the most important performance measures in the evaluation of the ATWRS. The analysis concludes with a discussion of the lessons learned about demonstration project transferability, the suitability of ATWRS as a model for deployment elsewhere, and the transferability of these lessons to other ITS projects.

HISTORY

The Fiscal Year 1991 Department of Transportation Appropriations Act included a provision requiring the Federal Highway Administration (FHWA) to study, develop and test highway systems to warn drivers of large trucks of the potential for rollover on highway ramps on the Capital Beltway in Virginia and Maryland. The impetus to the study was several costly and fatal accidents at Capital Beltway interchanges.

In 1993, the United States Department of Transportation (Federal Highway Administration – FHWA) completed a feasibility study that evaluated the application of systems to reduce truck rollover accidents on curved freeway ramps (USDOT, 1993). The report outlined and discussed countermeasures for preventing or reducing rollover accidents at highway interchanges. The findings of this study resulted in the installation of truck warning systems at three highway off-ramps, two in Virginia and one in Maryland. The ramp sites are located at I-495W/I-95S near Springfield, Virginia; I-495W/Route 123N near McLean, Virginia; and I-495E/I-95N near Beltsville, Maryland. These locations were identified in the initial feasibility study based on accident statistics from the Maryland State Highway Administration (MSHA) and the Virginia Department of Transportation (VDOT). Site selection was based on the geometric configuration of the ramp, truck volume on the ramp, the history of truck tipping on the ramp, and the characteristics of the area in terms of economic development, congestion and the number of vehicles traveling over the ramp.

FHWA participated in the identification and selection of the locations in conjunction with the Virginia and Maryland Departments of Transportation. A cooperative agreement between FHWA, VDOT, and MDOT outlined the terms for construction and maintenance of the system. Additional institutional cooperation provided for the collection and analysis of accident data. This effort involved the FHWA, the Maryland and Virginia State Patrols, local police jurisdictions, VDOT, and MDOT.

Identical warning systems, installed at each of the three locations, remain in operation today and data continue to be collected to evaluate the perfor-mance and effectiveness of the systems.

THE TECHNICAL SYSTEM

The essential components of the ATRWS include: an inroad detection/warn-ing system that identifies a truck and its relevant parameters (for example, speed and weight); and a warning device that consists of a flashing fiber-optic message sign and a static sign. Both are positioned before the curved ramp.

A roadside electronic controller receives the signal from the detectors, pro-cesses the information according to an algorithm that determines if the truck's speed may cause a rollover, and transmits a signal to activate the warning device if the truck's speed is greater than or equal to the rollover threshold speed.

Trucks are identified through a combination of inductive loops and piezo-electric sensors coupled with the controller that processes the data. A micro-wave-based radar beam functions as an "electric eye" to detect vehicle height and to establish whether trucks are above or below an established threshold height. The system has count station capability, but cannot identify truck type.

In addition to identifying the presence of trucks, the embedded inductive loop detectors and piezoelectric sensors work with the controller to determine truck speed. The truck's deceleration profile is determined by installing speed detection systems in at least two locations upstream of the curved ramp sec-tion. The truck rollover threshold is determined through the weigh-in-motion system which consists of the inductive loops and piezoelectric sensors. The system includes ten piezoelectric sensors and eight inductive loops.

The roadside electronic controller accepts the electrical inputs from the de-tection devices, processes the charges to identify trucks that exceed the rollover threshold and sends signals to activate the fiber-optic warning sign. The warn-ing sign message "TRUCKS REDUCE SPEED" only flashes if a truck is at or is exceeding the rollover threshold speed. The message is deactivated after the truck passes the sign. The fiber-optic sign is accompanied by a standard static warning sign featuring a tipping truck and the ramp advisory speed.

All of the information collected by the controller is automatically recorded by the roadside computer. State transportation officials can access this infor-mation from their offices via a cellular telephone connected to the controller. Both the computer and the cellular connection are located in the roadside control box beyond the shoulder guard rail.

The ATRWS is installed for two lanes, with signs on both sides of the ramps at the I-495W/I-95S and I-495E/I-95N intersections near Springfield, VA and Beltsville, MD. The system at I-495W/Route 123N near McLean, Virginia is installed for the single lane and the sign is only on one side of the ramp. All the systems were installed by International Road Dynamics, Canada in November 1993.

APPLICATION TO NATIONAL ITS GOALS

The primary purpose of the ATRWS is to prevent or reduce truck rollover accidents by warning drivers when their trucks will overturn unless they reduce speed. The operation of the ATRWS achieves this straightforward objective and also addresses six National ITS Goals (Euler *et al.*, 1995).

First, fewer truck rollover accidents on highway ramps will improve the safety of the nation's surface transportation system. The prevention of, or decrease in, truck rollover accidents not only improves highway safety for truck drivers, but also improves safety for motorists attempting to use the same interchanges and connecting highways.

Second, the ATRWS increases the operational efficiency and capacity of the surface transportation system. The prevention and/or reduction of truck rollover accidents reduces traffic delays caused by these accidents. This is a significant benefit considering that trucks overturning on exit ramps at interstate interchanges account for five of every 100 fatal truck accidents (USDOT, 1981).

Third, prevention and/or reduction of truck rollover accidents reduces energy and environmental costs associated with traffic congestion. For example, a March 1995 gasoline tanker crash on a Capital Beltway interchange blocked traffic for 12 hours and spilled 7000 gallons of fuel into a lake and stream south of the accident site (Hall and Tousignant, 1995). Prevention of this accident would have avoided the resulting energy and environmental costs, not to mention the delays experienced by many motorists.

Congestion reduction also meets the additional ITS National Goals: to enhance personal mobility, and to enhance the convenience and comfort of the surface transportation system. Although it is difficult to measure benefits from reduced accident rates, it is clear that truck rollover accidents inhibit the productivity and mobility of motorists.

ATRWS PROJECT GOALS

The specific ATRWS project goals were identified by examining the purpose of the project feasibility study. Although they are not explicitly referred to in the feasibility study, the project goals are implicit in the feasibility study and the first evaluation period report. The project goals derived from these assessments are:

- reduce truck rollover accidents on highway exit ramps;
- reduce associated congestion and delay resulting from truck rollover accidents;
- reduce accidents, injuries, fatalities, and property damage resulting from truck rollover accidents;
- operate a truck warning system funded by the federal government and maintained by state departments of transportation;
- install a cost-effective truck warning system.

ATRWS AND *EX POST* EVALUATION

It is important to examine the alternative to ATRWS that were considered at the time of project planning. The initial feasibility study considered several alternatives to the ATRWS including non-ITS ones. The study concluded that the ideal automatic warning system would warn a truck driver far enough upstream from a curved ramp to reduce speed to a level less than the threshold speed that would cause the truck to roll over on the curved ramp given the horizontal degree of curvature and super-elevation.

FHWA considered the continuation of the static sign as its base case. Although the static sign could be enhanced by flashing lights and/or a constantly flashing warning message, these signs had not proven effective in preventing truck tipping and it was found that drivers would not obey a constantly flashing sign to the same extent as a sign specifically intended for trucks exceeding the threshold speed.

The second option was the combination of the static sign and the inroad detection/warning system outlined above. This option could include the message sign warning trucks to reduce speed, or it could warn trucks by flashing the speed of the individual trucks on the message sign. The system could use only two vehicle variables: truck classification and speed, or the piezoelectric programmable classifier controller that also detects truck weight and height. This later variation would provide greater accuracy because of the additional input. However, a system based solely on truck classification and speed would be activated more frequently (and not always accurately) than a system that relied on additional information to assess the threshold for rollover.

Another alternative to embedded inductive loops or piezoelectric sensors is video imagery, a technology currently used for some freeway surveillance systems. This alternative was not selected because further development of video imaging technology was required before it could be used to reliably detect the necessary truck characteristics.

A final option would utilize on-board computer technology. At the start of each trip the driver would enter information on the vehicle configuration, cargo type, cargo weight, and load distribution in the truck's on-board computer. Each curved ramp with a history of rollover accidents, or with a combination of degree of curvature and super-elevation found to be associated with truck rollover would transmit the ramp geometrics from a transponder to the truck's on-board computer. Subsequently, the driver would receive a warning to reduce speed if there was a possibility of rollover. This system is referred to as an in-vehicle detection/warning system. However, all trucks would have to install the on-board system in order to achieve the goals of the project. This option was rejected because of the high cost to the trucking industry and the inability to assure widespread diffusion and adoption of on-board computers.

COST OF INSTALLATION, MAINTENANCE, AND OPERATION OF ATRWS

The initial estimate of installation and operator training costs for the ATRWS was $78 000 in 1993 (USDOT, 1993, p. 52). This cost included equipment, development, installation, training, operations, and maintenance. The cost of equipment and components included: a programmable classifier, a tracker narrow beam radar height sensor, inductive loop and piezoelectric sensors, a cabinet to contain the roadside controller, cable, the flashing signs, conduit and junction boxes, and discrete components. The installation cost also included installation of all equipment and components. The cost of operator training included manufacturer-provided training on the controller. The annual operation and maintenance costs were initially estimated at $3000. This cost included the cost of electrical power to support the system, periodic replacement of the inductive loops and piezoelectric sensors, and replacement of the lamps. The total cost of $81 000 for installation, operator training, and operations and maintenance was for a one-lane installation.

An updated estimate based on the final designs for the three ramps and actual bids from an equipment provider and an installer contractor led to a revised second estimate (ibid., 52). The second estimate for construction cost was:

Ramp 1 (I-495W/Rt. 123N, VA) $104 000
Ramp 2 (I-495W/I-95S, VA) $177 000
Ramp 3 (I-495E/I-95N, MD) $193 000

This estimate did not include several additional items that would later be added to the construction costs. These included a controller modification cost of

$22 500 to meet the requirements of the system. This was a one-time cost for the three projects as well as any future deployments. Additional expenses also included an approximately $5000 per site cost for system calibration, commissioning, and testing. An operator training charge of $2000 per state was also to be added to the total estimated cost. A final additional expense was for engineering design costs of $10 000 and $15 000 for single and dual-lane installations, respectively (ibid.). The total estimated design, construction and installation costs were :

Single-lane ramp	$121 000
Dual-lane ramp	$206 000

The system is expected to have a service life of 10 years. Annual maintenance and operation costs were estimated at about $1000 per year in this final estimate. However, the actual costs of design, construction and installation were significantly higher than the estimated costs. The actual construction cost per ramp as shown below represents a 27 percent increase over the estimated cost (USDOT, 1995).

Ramp 1 (I-495W/Rt. 123N, VA)	$149 542
Ramp 2 (I-495W/I-95S, VA)	$254 329
Ramp 3 (I-495E/I-95S, MD)	$240 845

These are the total costs with the exception of a $23 100 cost for controller modifications to the three systems, and a $2760 cost for system calibration, commissioning, and testing at the three sites.

The higher costs were due, in part, to the cost of bringing utilities to the system. This included the cost of connecting a telephone line and a power source to the site. The maintenance costs were also higher than estimated. For example, refurbishing one of the ramps cost $40 000. Some maintenance projects require total removal and replacement of the in-road sensors in addition to re-evaluation of the ramps after renovation. Changes in the ramp pavement can alter the threshold truck rollover speeds as a result of changes in the super-elevation of the curved ramp. After the maintenance project is completed, engineers must evaluate the curves and perform system adaptations as necessary. All of these factors contributed to the higher than estimated system deployment and maintenance costs.

COST-EFFECTIVENESS ANALYSIS

The decision to proceed with installation of the ATRWS at the three locations was based in part on an analysis of the cost-effectiveness of the system (USDOT, 1993, p. 53). This traditional analysis determined how many accidents would have to be eliminated by the system in order for it to equal the cost of the system. The analysis was based on estimates of truck rollover accident costs ranging from $13 274 to $15 470 per accident. These figures do not include the costs of potential fatalities or substantial infrastructural or environmental damage potentially resulting from hazardous cargo spills. The loss estimates for the ATRWS systems were based on the initial installation estimated costs plus a $1000 per year cost for maintenance over the 10-year life of the system.

The analysis concluded that a single-lane system would have to eliminate 6.55 accidents at an average cost of $20 000 per accident during the 10-year period. The two-lane system would have to eliminate 10.8 accidents at an average cost of $20 000 per accident were a 10-year life of the system. For accident costs of $100 000 per accident, 1.31 accidents would have to be eliminated on the one-lane system and 2.16 accidents would have to be eliminated on the two-lane system.

The cost-effectiveness of the system is dependent on preventing high-cost rollover accidents. Records of pre-installation accident frequency and costs were assessed in the FHWA Feasibility Study to measure post-installation effectiveness. The analysis of the accident data showed that the system could be cost-effective for ramps with a history of truck rollover accidents at least once every five years.

The *First Evaluation Period Summary Report* and the *Final Report* also evaluated the cost effectiveness of the Automatic Truck Rollover Warning System using installation costs and accident data (USDOT, 1995, 1998). No truck rollover accidents occurred at any of the three sites during the first or final evaluation periods, nor has any occurred through July of 2000. Consequently, the system appears to be cost-effective based on the previous analysis and subsequent history.

PERFORMANCE CHARACTERISTICS

The performance characteristics of the ATRWS are dependent on several factors. First, successful operation of the system is dependent on the success and accuracy of the in-road detection/warning system. Second, performance is dependent on the ability of the sensor/detection stations, the controller and the fiber-optic warning sign to operate as a system. Third, acceptable ATRWS performance depends on labor-intensive maintenance and re-installation procedures for the in-pavement components that may include the cost of ramp reconstruction or resurfacing. Performance information collected at the site is sent to VDOT headquarters through the cellular link to the site. This is the primary means of confirming that a system is operational and acquiring evaluation data.

The *First Evaluation Period Summary Report* and the *Final Report* included a number of traditional evaluation measures that were used to assess system performance (USDOT, 1995, 1998). Calibration tests evaluated system detection accuracy to insure that all components worked and the systems functioned according to their design. Both visual and electronic observations were compared to verify the measurements of the system. A final calibration test compared laser speed gun measurement for individual weigh-in-motion stations.

A total systems evaluation analyzed the collected data to determine the warning system's effectiveness on speed reduction of high speed trucks. This evaluation considered whether the fiber optic message sign was activated by high speed trucks and whether these trucks reduced speed after passing the message sign.

The evaluation section of the reports included general and ramp-specific guidelines for future installations and changes to the existing systems. Most of these recommendations focused on changes in system design and construction, and changes in future evaluation methods (USDOT, 1998).

INSTITUTIONAL EVALUATION ISSUES FROM APPLICATION OF THE CHECKLIST

The selection, installation and operation of the Automatic Truck Rollover Warning System provides insight into institutional considerations relevant to future ITS projects. These insights are discussed below and also appear in the evaluation checklist summary in the Appendix to this chapter.

One of the most obvious institutional findings is the necessity of intergovernmental cooperation. Cooperation between the Virginia and Maryland Departments of Transportation and the US Department of Transportation was essential for the successful installations of the system and remains essential for ongoing operations. For example, although the federal government funded construction of the system, the state governments are responsible for maintenance. The state departments of transportation relied on the accuracy of maintenance estimates for budgeting and could not have anticipated the higher-than-expected maintenance costs.

Similarly, multiple jurisdictions are involved in accident analysis. Although the relevant state patrols respond to accidents on the Capital Beltway, there is a jurisdictional question regarding whether the state patrol or the local police agency is responsible for responding to and reporting accidents on highway ramps. Accurate reporting requires both jurisdictions to code their reports using the same format or template. This presented a problem because the reporting system for highway ramp accidents previously relied on the written description of the accident and did not include a report category for the number or location of the highway ramp.

Potential liability resulting from the failure of the ATRWS system provides another example of the importance of considering institutional issues prior to installation of an ITS project. In this case, the project managers chose to maintain the traditional static sign featuring a graphical representation of a tipping truck and the recommended safe speed. The static sign was installed in addition to the new system's flashing sign in order to counter potential liability issues. The static sign provides protection against potential failure of the ATRWS; this protection is further enhanced by the remote dialing system which allows daily remote analysis of data collected by the ATRWS to scan for system errors. The cellular connection also offers convenience and cost savings by eliminating the need for daily spot checks of the system at the ramp site.

Privacy concerns could present another institutional challenge for project managers, although they have not influenced system design or operation to date. For example, the operating speed of the trucks on the ramps and other data collected by the controller are public information because the project is publicly funded. Although no one has requested this information, its availability could present unforeseen challenges for the state departments of transportation in terms of allowing the public to access the data collected at the site. Privacy could also become a matter of concern for the vehicle operators. Although ATRWS signs do not specifically post speed, other vehicles are alerted to the trucks' excessive speeds by activation of the sign. Despite the privacy issues raised by the ATRWS, the system currently in place is less invasive than the rejected alternative that relied on on-board computers.

Lastly, this project has important institutional lessons for other ITS projects. Although the initial funding for installation and project evaluation was provided by the federal government, state governments will be responsible for future increases in the number of deployment sites. Proliferation of the technology will be dependent on a number of constraints. Additional sites will remain dependent on state funding availability and the expressed desire for installation of the ATRWS by local jurisdictions. These institutional constraints are further complicated by the physical limitations of potential sites. In order for the ATRWS to function effectively, the number of potential sites is limited by ramp geometry. Ramp geometry must result in a specific mix of ramp length, curvature, super-elevation, and grade for it to benefit from ATRWS.

The matrix of evaluation criteria (see Appendix) addresses these and other institutional issues. Although the traditional evaluation criteria should be considered before an ITS project begins, these institutional issues illustrate the importance of considering a broader range of criteria before a project begins.

NON-TRADITIONAL EVALUATION

This section discusses evaluation-related findings for the Automatic Truck Rollover Warning System. Although numerous traditional evaluation efforts have been pursued, the findings discussed below result from considering non-traditional evaluation criteria.

In the case of the ATRWS, direct methods of performance measurement may not yield the most useful information. For example, some ramps may have experienced zero truck tipping accidents before installation and zero accidents after installation. In this instance it is difficult to attribute the success of the ATRWS to the continuation of an absence of truck accidents. Instead, the systems' benefits may become apparent as traffic volumes, speeds or driver behavior change. These changes may contribute to the absence of truck tipping accidents, but may not be apparent in the accident reports used in traditional evaluation reports. The evaluation lesson to be drawn from this example is that for some ITS projects, new methods of performance measurement should be included in evaluations.

One of the potentially limiting factors for future deployment of the ATRWS is that the cost of the system is high. Unlike traditional highway infrastructure, there is a small market for ITS systems such as ATRWS[1] and an even smaller supplier base; for example, there is currently only one supplier for the total ATRWS system, International Road Design of Canada (IRD). These supply and demand conditions inflate the price of the system. The unexpected project requirements and the necessary coordination of contractors through the state bidding process also increased the base costs for the ATRWS. Another unforeseen cost resulted from utility companies that experienced difficulty accessing the sites. The nature of the project required more than the usual number of sensor stations to ensure adequate data were produced and collected, thereby further increasing project costs. The evaluation lesson drawn from these cost factors is that the value of ITS demonstration/operational tests as models may not be as high as is typically thought. Their anticipated demonstration value may need to be reduced.

One of the specific system components contributing to the high cost of the ATRWS is the cellular links. Establishing cellular links was more difficult and expensive than expected. Part of the expense resulted from the initial billing period. The project managers had not anticipated that the cellular link would result in continuous redial to the state department of transportation office on weekends. As no employees were present to accept the calls on weekends, the system continuously re-dialed the office. This resulted in an unexpectedly high phone bill for the first month. The problem was then corrected by reconfiguring the system. However, configuring the system for automatic dialing to the department of transportation headquarters for daily data dumping proved to be more difficult than expected and resulted in further additional expense. The evaluation lesson drawn from these difficulties and unanticipated expenses is that in evaluating ITS projects, the costs, capabilities and uncertainties of complementary technologies (for example, cellular communications) may affect the evaluation, and can be expected to create unexpected contingencies and, therefore, cost.

In addition to the generally high installation costs, the total final costs were significantly higher than estimated. Total costs increased from the feasibility study to the first evaluation period; and increased again from the first evaluation period to subsequent cost estimates. The evaluation lesson to be drawn from theses increases is that initial estimates will need to be adjusted as the project proceeds to reflect cost escalation.

The importance of project expenses will undoubtedly influence the future deployment of the ATRWS. The lessons that can be drawn from these specific examples are transferable to other high-cost ITS operational tests. Because technologies such as ATRWS are in the early development stage, evaluations should assume that overall costs will be higher than expected until full-fledged user markets develop.

The formal FHWA assessment of the ATRWS was a traditional evaluation effort, relying primarily on measures of the absence or reduction of truck tipping accidents. This effort did not include feedback from motor carriers and truck drivers. This feedback, which is necessary to provide a complete evaluation, could be obtained through the use of one or more methods. Possible methods include: user groups or focus groups; expert opinion; consultation with interest group organizations; and/or survey research. The lesson that can be drawn from this finding is that evaluation methods for some ITS projects should include techniques to reveal user attitudes in addition to traditional evaluation methods.

The chosen evaluation method should also consider the effects of highway/infrastructure maintenance. Traditional highway infrastructure is maintenance intensive, and it often does not have any embedded electronics. In the case of the ATRWS, the unexpected construction and renovation of ramps significantly increased costs, which may ultimately lead to negative attitudes toward the ATRWS. Furthermore, reinstallation of the ATRWS after construction is dependent on available contractors and state government bidding procedures. The lesson that can be drawn from this is that ITS evaluations should be structured to illustrate how well a given project fits into a complex, maintenance-intensive infrastructure network.

The operation of the ATRWS also reveals that well-designed, but possibly unnecessary evaluation hardware may dampen enthusiasm for future deployments of the system. The ATRWS system includes voluminous data and performance information, and collection capabilities (for example, multiple in-pavement loop detectors and the microwave-based radar sensors). However, future deployments will not require these expensive data collection systems. The evaluation lesson that can be drawn from this is that efforts to provide hardware and software for rigorous ITS evaluation may actually complicate evaluation efforts that could be valid for other sites.

The decision to select the ATRWS over the on-board computer warning system also has important implications for other ITS technologies. Although the use of the ATRWS in conjunction with on-board computers may become a more feasible option in the future as the technology matures, the decision makers chose to adopt the relatively simpler sign-based ATRWS, which could be deployed without delay. Future ITS evaluations should recognize that more rapid deployment of simpler technologies may be valuable, and therefore more weight should be given to the "time" criterion the evaluation. In this case, the ATRWS was deployed immediately and began to meet the project goals even though more sophisticated technology was available.

Finally, while traditional evaluation methods may assume that a lack of truck tipping accidents stem from installation of the ATRWS, other factors could contribute to a decrease in accidents. The assumption that a lack of truck tipping represents system success is a possible instance of faulty causation. Other factors that could contribute to a lack of truck tipping accidents include: enhanced speed enforcement; variations in freight characteristics; and increased driver familiarity with highway design; and new breakthrough technology, such as widespread adoption of on-board computers. Ultimately, ITS evaluations should be careful to avoid reliance on extraneous variables or conclusions based on faulty causation.

SUMMARY OF EVALUATION LESSONS

- For some ITS projects, new methods of performance measurement should be included in evaluations.
- The value of ITS demonstration/operational tests as models may not be as high as is typically thought. Their anticipated demonstration value may need to be reduced.
- Many initial benefit-cost analyses should be adjusted to reflect likely cost escalation.

- Since these technologies are in the early development stages, evaluations should assume costs will be high, and higher than expected, until full-fledged user markets develop.
- In evaluating ITS projects, the costs, capabilities and uncertainties of complementary technologies (for example, cellular communications) may affect the evaluation, possibly providing unexpected contingencies.
- For some ITS projects, evaluation methods that include techniques to reveal user attitudes would add valuable insight into the utility of a project.
- ITS evaluations should be structured to illustrate how well a given project fits into a complex, maintenance-intensive infrastructure network.
- Efforts to provide hardware and software for rigorous ITS evaluation may actually complicate evaluation efforts that can be valid for other sites.
- ITS evaluations should recognize that more rapid deployment of simpler technologies may be valuable, and therefore more weight should be given to the "time" criteria in the evaluation.
- ITS evaluations should be careful to avoid extraneous variables and acknowledge the possibility of conclusions based on faulty causation.

CONCLUSION

Construction, maintenance and operation of the Automatic Truck Rollover Warning System progressed as planned and included a traditional evaluation effort. Although this traditional evaluation effort provided critical information for assessing the effectiveness of the system, this assessment reveals the need for consideration of other non-traditional evaluation considerations to assess indirect and unexpected effects, transferability, and the project's suitability as a model. The evaluation-related findings of the study are applicable to other ITS projects (for example, issues of indirect effects, transferability, and suitability as a model for deployment elsewhere). The multi-criteria assessment of the ATRWS reveals the diversity of ITS projects and illustrates the specific goals and accomplishments of this system.

Specific project related lessons are:

- unexpectedly high costs and site access problems in terms of hardware and utilities installation;
- difficulties with the automated cellular phone system and its higher than expected costs;
- internal staffing uncertainties;
- rebuild problems related to pavement reconstruction and maintenance;

- in-pavement versus roadside equipment lessons;
- high prevailing speeds and driver behavior's effects on compliance;
- unanticipated downtime;
- effects of system downtime on driver acceptance;
- effect of high evaluation costs on enthusiasm for other sites;
- as an operational test, its effects upon a huge multi-mission agency may be too limited and insufficient to build institutional and technical support for future deployments unless significant cost reductions are seen.

APPENDIX: ITS EVALUATION CHECKLIST EXPLANATIONS

For the ATRWS case, the most important performance measures in the checklist (see Appendix to Chapter 1) include: "Congestion and delay"; "Safety and accidents"; and "Cost relative to other options". These categories contain the most important effects because they have the greatest impact on the project goals. These and other categories in the checklist are discussed in greater detail below. The checklist format allows the evaluator to ascertain the most critical effects and measures.

Congestion and Delay

This category is one of the most relevant and is a directly applicable measure of the ATRWS. An absence of truck rollovers could lead to an increase in peak-hour VMT and a decrease in vehicle-hours of delay, peak/non-peak travel times, and average length of queue. This category can be measured by comparing current post-installation data regarding these indicators to similar data recorded prior to installation of the ATRWS. A summary assessment of specific indicators appears in Table 5A.1.

Capacity and Mobility

This category is similar to congestion and delay in that the positive effects of these indicators result from a lack of truck rollover accidents. This category can also be operationalized by comparing data for each of the indicators – pre and post-installation. It is important to note that these effects may experience variable decreases in response to on-site maintenance which closes one or more lanes of the highway ramp.

Both of these categories also depend on reliable accident data which clearly attribute the cause of specific accidents to truck rollovers on specific highway ramps.

Table 5A.1   Automatic Truck Rollover Warning System checklist:
Congestion and delay

Goal[a]	Indicator	Effect	Comment
2	Peak-hour VMT	Increase	Most important benefits
2	Vehicle-hours of delay	Decrease	Result of reduced accidents and congestion
2	Volume/capacity ratio	Increase	As ramp accidents and queues are reduced
2	Peak/non-peak travel times	Decrease	Fewer accidents and reduced congestion
2	Average speed (e.g. PMT/P-Hr)	No effect	Measured statistically
2	Average length of queue	Decrease	As accidents decrease

Note: [a] Project goals: Column 2 of Checklist in Appendix Table 5A1. Reduce truck rollover accidents on highway exit ramps; 2. Reduce associated congestion and delay resulting from truck rollover accidents; 3. Reduce accidents, injuries, fatalities, and property damage resulting from truck rollover accidents; 4. Operate a truck warning system funded by the federal government and maintained by state departments of transportation; 5. Install a cost-effective truck warning system.

Table 5A.2   Automatic Truck Rollover Warning System checklist: Capacity
and mobility

Goal	Indicator	Effect	Comment
2	Ease of personal mobility	Increase	Result of fewer accidents and reduced congestion
2	System accessibility	Increase	Result of fewer accidents and reduced congestion
2	Network or connectivity implications	Increase	Less delay, more certainty
2	Daily VMT	Increase	Result of fewer accidents and reduced congestion

Safety and Accidents

The decrease in deaths, injuries and property damage can be directly attributed to a lack of truck rollover accidents. Measuring these indicators requires
access to accurate accident reports. The most direct measure of the first two
indicators is to compare pre- and post-installation data. These data are included in the Beltway Accident Database.

*Table 5A.3   Automatic Truck Rollover Warning System checklist: Safety
              and accidents*

Goal	Indicator	Effect	Comment
3	Deaths	No effect	No recorded deaths preceding installation
3	Injuries	Decrease	Truck tipping prevented
3	Property Damage	Decrease	Truck tipping prevented

Cost Relative to Estimates

This effect is applicable and relevant because it could be a key factor in the
decision making process for other jurisdictions considering installation of the
ATRWS. Total costs exceeded initial estimates. Although future sites could
circumvent some of the higher costs, other costs (for example, installation of
in-pavement sensors) would remain high.

*Table 5A.4   Automatic Truck Rollover Warning System checklist: Cost
              relative to other options*

Goal	Indicator	Effect	Comment
4	Construction	Higher cost	Cost exceeded estimates
4	Operations	Higher cost	Unforeseen telecommunication expenses
4	Maintenance	Higher cost	In-pavement installation difficult and costly

Financial and Fiscal Outcomes

This category could be more relevant for projects in other jurisdictions. Budgetary sufficiency is not particularly relevant to this case study, as it is a federally funded project. The argument can be made that private benefits are increased as a result of decreased congestion and delay. Operationalization of

this indicator depends on reliable accident data and literature regarding the linkage of increased private profits and decreased congestion and delay.

Table 5A.5   Automatic Truck Rollover Warning System checklist: Financial and fiscal outcomes

Goal	Indicator	Effect	Comment
4	Budgetary sufficiency	None	Not funded w/state funds
4	Inter-governmental transfers	Mostly FHWA funds	FHWA funds/VDOT and MDOT fund maintenance
4	Private profits	Increase	Decreased congestion and delay

Environmental Outcomes

Decreases in ambient emissions levels and ground level ozone result from decreased congestion and delay. Measurement of these indicators relies on accident data and literature regarding the effect of congestion and delay on emissions and ozone levels. As the ATRWS requires minimal modification to the existing highway ramp and surrounding areas, there are no subsequent effects on soil erosion or wetlands and natural habitats.

Table 5A.6   Automatic Truck Rollover Warning System checklist: Environmental outcomes

Goal	Indicator	Effect	Comment
—	Total emissions	Slight decrease	Reduced congestion-related idle time
—	Ground level ozone	Slight decrease	Reduced congestion-related idle time
4	Energy consumption	Slight decrease	Reduced congestion-related idle time
—	Hazardous materials impact	No effect	No previous haz. mat. accidents on these ramps
—	Soil erosion	No effect	Minimal construction
—	Wetlands and habitat	No effect	Minimal construction

Economic Outcomes

Measurement of this category depends on the existence of literature revealing a linkage between reduced congestion and delay and economic outcomes. Gross regional product, employment, per capita income, land values and regional competitiveness could increase as a result of fewer truck accidents, but this is a difficult argument to make considering the numerous factors involved in highway access and congestion.

Table 5A.7   Automatic Truck Rollover Warning System checklist: Economic outcomes

Goal	Indicator	Effect	Comment
1	Employment	Indirect effect	As accidents and congestion are reduced
—	Per capita income	Indirect effect	As accidents and congestion are reduced
—	Land values	No effect	Not applicable
—	Regional competitiveness	Increase	If deployment is widespread

Equity Outcomes

The ATRWS has no effect on affordability, regional access disparity, or handicap access because these indicators do not change as a result of installation of the system.

Table 5A.8   Automatic Truck Rollover Warning System checklist: Equity outcomes

Goal	Indicator	Effect	Comment
—	Affordability	No effect	Not applicable
—	Regional access disparity	No effect	Not applicable
—	Handicap access	No effect	Not applicable

Institutional Outcomes

The institutional outcomes are relevant. The public sector does not experience increased liability because VDOT maintained the static sign in addition to the variable message sign. The static sign was maintained to serve as a precaution against potential variable message sign failure. Privacy has not become an issue at this time, but the potential exists for this issue to develop. The ATRWS records the time each vehicle crosses the ramp and measures the speed of each vehicle. The availability of this public information for lawsuits filed against truck drivers who fail to heed speed warnings and roll over on the ramp could become an issue. Installation and testing of the system required coordination of multiple institutions working together. Finally, intergovernmental cooperation is essential to the operation and maintenance of this project.

Table 5A.9 Automatic Truck Rollover Warning System checklist: Institutional outcomes

Goal	Indicator	Effect	Comment
4	Flexibility, adapts	None	Dependent on fixed infrastructure
—	Regional significance	Yes	If it reflects regional ITS strategy
—	Synergy	Yes	Necessary for installation and testing
4	Intergovernmental friction/ cooperation	Yes	Early tests are joint efforts; future deployment; no effects
—	Community involvement and preferences	No effect	Not applicable
—	Ownership changes	No	Built on public property
—	Public–private partners	No	No private partners
—	Liability	No	Does not increase public sector liability/speed was already posted
—	Privacy	No effect	May increase demand for public records

Spatial Outcomes

The spatial outcomes are not particularly relevant to the ATRWS. The project had no effect on land use, land values or residential proximity issues. Although the ATRWS did not have an effect on the design of the exit ramps, infrastructure design is a key factor for other jurisdictions considering installation of the ATRWS.

Table 5A.10   Automatic Truck Rollover Warning System checklist: Spatial outcomes

Goal	Indicator	Effect	Comment
—	Land use implications	No effect	Not applicable
—	Residential proximity issues	No effect	Built on existing roads
—	Infrastructure design or geometry impact	No effect	ATRWS adapted to fit current infrastructure; future ramp designs may have to differ

Demonstration Outcomes

The ATRWS has potential transferability, but, again, other jurisdictions will be limited by their exit ramp configurations. The project also has questionable value as a model because the system included additional reporting and monitoring devices which will not be necessary in future deployments.

Table 5A.11   Automatic Truck Rollover Warning System checklist: Demonstration outcomes

Goal	Indicator	Effect	Comment
—	Transferability	Yes	To other jurisdictions but not to every ramp
—	Value as a model	Poor	High and unexpected costs, excessive monitoring, hardware installation
5	Suitability	Uncertain	No direct linkage between ATRWS and reduction of accidents; drivers may not heed signs; unreliability

NOTE

1. Illustrative of this is that as of July 2000, no additional ATRWS was created in the Washington region, although one additional system is planned for deployment in the coming year.

REFERENCES

Euler, Gary W. and Robertson, H. Douglas (eds) (1995), *National ITS Program Plan,* Vol. I, Washington, DC: ITS America.

Hall, Charles W. and Tousignant, Marylou (1995), "Gas tanker crash sparks safety fears", *Washington Post* 18 March, B1.

US Department of Transportation. Federal Highway Administration (1981), *The Effect of Truck Size and Weight on Accident Experience and Traffic Operations* (FHWA-RD-80-137).

—— (1993), *Feasibility of an Automatic Truck Rollover Warning System* (FHWA-RD-93-039).

—— (1995), *Evaluation of a Prototype ATRWS Truck Warning System: First Evaluation Period Summary Report*, submitted to Turner-Fairbank Highway Research Center by Bellomo-McGee, Inc., April 25.

—— *Evaluation of Prototype Automatic Truck Rollover Warning Systems: Final Report*, submitted to Turner-Fairbank Highway Research Center by Bellomo-McGee, Inc., Jan. (FHWA-RD-97-124).

Vallete, G., McGee, H., Sanders, J. and Enger, D. (1981), *The Effect of Truck Size and Weight on Accident Experience and Traffic Operations* (FHWA-RD-80-137), Washington, DC: Federal Highway Administration, July.

6. The Montgomery County Advanced Transportation Management System

Hans Klein

OVERVIEW

This study examines the Advanced Transportation Management System (ATMS) in Montgomery County in the state of Maryland. Montgomery County's ATMS is one of the largest implementations of ITS in the United States today, combining a large-scale traffic management system with a variety of information diffusion services. The ATMS predates the national ITS program by over 10 years, so it can offer useful lessons about barriers to both technical functionality and the dynamics of system development (Knopp and Lynch, 1996).

THE TECHNOLOGY: CONTEXT, FUNCTIONALITY, HISTORY

Montgomery County is a densely settled suburban county (population 800 000) outside of Washington, DC. The county has an extensive network of freeways, arterial roads, and surface streets. Figure 1.1 in Chapter 1 shows the size and location of the county relative to the District of Columbia, in addition to the major roads in the county.

Between 1978 and 2000 the Montgomery County Department of Transportation (MCDOT) installed and continuously expanded its Advanced Transportation Management System. The system has been designed largely by the MCDOT staff, with much of the initiative coming from the Chief of the Transportation Systems Management Section, Mr Gene Donaldson. As one of the most advanced implementations of ITS in the United States, Montgomery County's ATMS offers insights into the challenges of developing and evaluating ITS. Descriptions of other more recent initiatives are found in Kaiser (1999); Mitchell (1998); Pittenger (1997); Florida Department of Transportation

(1995); Havinoviski *et al.* (1996); Jacobson and Sumner (1995); and Kerr *et al.* (1997).

The Technology

The Montgomery County ATMS has a 20-year history and continues to develop even today. For the purpose of analyzing the technology and considering its evaluation, this section of the chapter uses a snapshot of the system as it existed in the late 1990s (see Adeyinka, 1997; Donaldson, 1992). A later part explores the history of the ATMS, examining the system's dynamics. This includes:

- surveillance;
- data processing;
- traffic control;
- information diffusion.

Surveillance

The ATMS employs a variety of surveillance and detection systems to gather information on current traffic conditions in the network. These data are augmented by information from other organizations that monitor traffic conditions. The surveillance technologies include:

- in-pavement detectors (Berka and Lall, 1998);
- video cameras (Awadallah and Habesch, 1998);
- aerial surveillance.

The system also includes GPS (global positioning system) based automatic vehicle location (AVL) systems (Watje *et al.*, 1994; Hill *et al.*, 1996).

In-pavement Detectors

There are of two kinds of in-pavement detectors: sampling detectors that measure traffic volume, registering both the presence of a vehicle and the speed at which it travels (volume and occupancy). The Montgomery County ATMS has deployed roughly 2000–3000 sampling detectors and also some 8000–10 000 presence detectors, which are simpler, merely registering the presence of a vehicle (usually at a signalized intersection).

Video Cameras

Video surveillance is performed by cameras located near important sites on the road network. Equipped with remote-controlled telephoto lenses, the cameras can be pre-programmed to pan and zoom in on strategic locations on the road network at certain times of day. For non-routine operation, such as incident management, cameras can also be operated by staff from the traffic management center.[1] The ATMS currently has nearly 80 cameras in operation.

Aerial Surveillance

Aerial surveillance is performed by MCDOT's leased airplane. The airplane provides both visual and video surveillance. Airplane operators both radio down descriptions of current traffic conditions and download live video images. The use of aerial surveillance has proven to be one of the most useful technologies for monitoring the status of traffic on the system.

Information is also generated by specially equipped transit vehicles. Montgomery County's buses are equipped with GPS systems which provide information on current location as buses proceed through the transportation network.

In addition to its own surveillance technology, the ATMS receives and integrates traffic information from other organizations.[2] MCDOT maintains direct radio and telephone contact with police and fire departments. It also maintains direct communications with the Maryland State Highway Administration's statewide operations center. It has routine contact with commercial traffic reporting services (Metro Traffic and Shadow Traffic). The ATMS also receives information from its Traffic Watch Team – a team of volunteer citizens who have been trained in how to give accurate telephone reports to MCDOT. Table 6.1 lists the surveillance capabilities of the system.

The second function of the ATMS is data processing.[3] Data processing occurs at the MCDOT's Transportation Management Center (TMC) located in Rockville, Maryland. The TMC receives data over a variety of communication media, integrates it, analyzes it, and computes overall control strategies. Inside the TMC, video monitors show television images of key intersections, and computer graphics workstations display the current state of traffic signals. Many control strategies are automated, so that signal timing adjusts automatically in reaction to current traffic conditions. However, staff also formulate *ad hoc* strategies in response to particular incidents.

The third function of the ATMS, traffic control, is largely achieved through traffic signals.[4] The county maintains more than 600 signals, all of which are integrated into the ATMS. Their timing cycles are adjusted by the central computer in order to maintain traffic efficiency under changing conditions. Cycles are also changed to give priority to transit buses running behind schedule. In case of incidents, operators manually adjust signals.

Table 6.1   Surveillance capabilities of the ATMS

Devices	Quantity
Sampling loop detectors	Approx. 1000
Presence loop detectors	Approx. 10 000
Video	80
Aerial surveillance	1 full-time airplane
GPS-equipped transit vehicles	200 buses
Other agencies	Police and Fire
Traffic information companies	Metro Traffic, Shadow Traffic
Telephone	Citizen volunteers (1996)

Table 6.2   Information diffusion

Diffusion media
Cable television (Montgomery County Cable TV–Channel 55)
Travel advisory radio system (TARS)
Communication with other agencies
Video link to local television news stations
Internet
Kiosks
Variable message signs (6 portable)
Traveler advisory telephone systems (301-217-8800)

Lastly, the ATMS employs a broad array of media for information diffusion (see Table 6.2), sharing information with road users and outside organizations so that better informed travel-related decisions can be made.

Information from the ATMS is broadcast over a variety of mass media. Direct video links connect the traffic control center to local television news rooms; the staff can provide live traffic reports directly from the control center. Additionally, traveler advisory radio messages can be recorded by TMC staff at the TMC radio broadcast. The system also connects to the county government's cable access television channel. Images from roadside video cameras can be fed onto the cable system, supplemented by text and graphics information about current transit and traffic conditions.

There is also an Internet web site featuring the same graphics, data, and video displays available to TMC operators.[5] Internet users have access to the same information that is available to the system operators themselves. Information kiosks are also planned which would give access to information at key locations.

One common information diffusion technology, variable message signs (see Chapter 4), have not been used on the system. However, these are planned for the future (fixed and portable). Table 6.2 lists the various information diffusion technologies (Sussman, 1999; Hyman and Miller, 1996; Mohaddes, 1995).

The ATMS is more than a technical system: it is also an inter-organizational system. Coordination among different agencies in Montgomery Country is an important aspect of the ATMS. A Transportation Operations and Incident Management Team unites representatives from the county's police, fire and rescue, environmental protection, and transportation departments. This group meets monthly to plan responses to major events that would impact transportation (for example, sporting events, major accidents.)

Organizational integration has advanced most between the transportation department and the transit department. Not only do they share information from data collection and traffic control, the two organizations recently acquired adjacent office space in the building. Increased technical connectivity led to the decision to pursue greater organizational connectivity.

Over the 20 years of its existence, the total investment in the ATMS has been approximately $10 million. With expected investment of another $15 million in the coming years, the total investment in the ATMS should come to $25 million. Although state and federal funds have been used, county funds accounted for approximately 90 percent of total investment through to the late 1990s. Despite recent increases in state and federal participation, the county share of the total $25 million system will still be approximately 80 percent.

History of the Montgomery County ATMS

The Montgomery County ATMS just described is the product of nearly 20 years of work by the staff of the county's Department of Transportation. The history of system development reveals a process of steady, incremental growth from the original simple system.

The original system was proposed and designed by an outside consultant working for a real estate developer. In the late 1970s, a developer approached Montgomery County with a proposal for a large enclosed regional shopping mall. As part of that development project, 10 traffic signals in the area would be integrated into a computer controlled signal system. This proposal was accepted, and in 1978 a vendor was selected for the system.

Implementing the first system placed extraordinary demands on MCDOT staff. The system was originally housed in a small storage room in a county firehouse. Considerable overtime was required of staff to work out the problems in the system. Much of the extra work required to launch the system was performed by one individual, Gene Donaldson.

The technology used in the ATMS was initially relatively simple, but incrementally became more advanced. Initially the system operated without adaptive control; instead it employed executing control based on stored time-of-day routines. Only 10 traffic signals were included in the original system. However, every few years MCDOT staff made incremental improvements. In 1981, system parameters were set for adaptive traffic control so that traffic signal cycles could be modified in response to changes in traffic conditions.

Although the technology was not in itself innovative, its application demanded innovation in MCDOT. The staff had to learn new skills, and MCDOT's office space had to be reconfigured in order to accommodate the new technology. As the staff developed the necessary skills and as they acquired resources for growth, they developed the system further. In 1981 the central computer was moved out of the firehouse and into MCDOT's offices in the newly constructed county office building in Rockville. Through their extraordinary efforts, MCDOT staff became expert in the use of ATMS technology.

In 1981 planning began for an expanded system. In 1983, a larger computer with 256K of memory was purchased. A total of 50 traffic signals were put under computer control. Expansion of the system grew steadily, so that by 1986 some 200 traffic signals had been integrated into the system.

As demand grew, MCDOT staff designed an open system. Proprietary technologies were avoided so that later additions to the system would be easy. By maintaining an open technology, staff could integrate emerging technologies into the existing system.

The computer expertise developed by the staff also stimulated broader applications of computer technology within MCDOT. The same computers used for traffic control were soon used for administrative and engineering tasks. Staff wrote programs for preparing project cost reports, calculating signal timings, and producing left-turn phasing warranting analyses. Packaged software was acquired for word processing, spreadsheets, and databases.

This early experience lead to broad organizational expertise and commitment to computer applications. As an organization, MCDOT was transformed by these experiences. It developed internal expertise and became an advocate for further development. The goals of the organization changed through the experience of developing the first system. MCDOT had developed commitment to the new technology.

In 1989, MCDOT made another major change by using aerial surveillance to gather information. MCDOT staff and police staff began flights over county roads in order to identify congestion and incidents and to manage responses. Aerial surveillance greatly decreased the time needed to identify incidents and evaluate their nature and severity. This aerial view also provided an overview of the road network as a whole, allowing for a more systemic understanding of traffic behavior. Information provided by aerial surveillance could be integrated into the ATMS to manage traffic in response to major incidents.

In 1992, an additional series of new initiatives began. First, a county-wide ATMS was approved and $2 million was budgeted for the first phases of system deployment. The full ATMS would cost $25 million. Phases I and II of the ATMS would include the installation of 200 video cameras mounted along roads. At about the same time the FHWA provided funding for integrating video into the aerial surveillance. FHWA funds were met equally by county funds. Finally, a $1 million federal grant was won to use GPS systems to track the position of buses. This would be matched by $1.5 million of county funds and $2.5 million of state funds. The ATMS was suddenly growing on many different fronts.

At the heart of the enlarged system is a Transportation Management Center (TMC). The TMC became the central control point for the expanded system. The computers at the center automatically received traffic sensor inputs and adjusted the traffic signals accordingly. More sophisticated tasks, such as responding to incidents, would be performed by the staff. As the system grew, the importance of this central node increased proportionately.

In its early years, the ATMS stimulated organizational change within MCDOT, and, as the system expanded, changes appeared in other public agencies in Montgomery County.

MCDOT and the county police had already begun cooperating in the operation of the airplane. Subsequently, these two organizations joined with others to form the Transportation Operations and Incident Management Team. The Team meets on a monthly basis to plan coordinated responses to major incidents that impact the transportation system. The integration of transit vehicles into the ATMS also led to closer organizational ties with the transit agency. MCDOT was interacting regularly with police, fire, rescue, and transit agencies. Inter-organizational networks were forming.

The networks bore additional fruit. In the mid-1990s the police unit responsible for transportation activities reorganized to make better use of the ATM system. The new Traffic Enforcement and Accident Management police section would collaborate closely with MCDOT. In addition, the police plan to convert an old Greyhound bus into a "command bus" which will serve as a mobile information center, receiving aerial video download and other information so that police officers at an emergency site can make better decisions.

MCDOT's technical capabilities and commitment continued to produce new initiatives. In the mid-1990s the ATMS was linked to additional diffusion media. The system was integrated with the local cable television network to show real-time images of traffic conditions on television. Video links were established with local television broadcast stations. These features allow county residents to make informed transportation decisions while still at home. Travelers' advisory radio capabilities were also expanded. Mobile telephone was encouraged as a source of traffic information. Diffusion of traffic and transit information over the Internet has become a reality, with a web page displaying traffic information. The system benefited from economies of scope, as it increasingly connected with other communications media and other organizations to realize new functions at little additional cost.

In summary, the history of Montgomery County's ATMS is a story of steady activity and incremental implementation of new technologies. It began with an early contingent event: the proposal to computerize 10 traffic signals around a shopping mall. That first, simple system planted the seed which has continued to grow for many years.

The original technology was not particularly advanced. The original system – 10 signals operating on a fixed schedule – did not constitute advanced technology, not even in 1979. Yet the *implementation* of that technology was a major innovation, for it required extraordinary efforts by MCDOT staff to integrate it into their organization. Development was carried forward by the personal commitment and leadership of the staff. By beginning with a reliable technology, staff could focus their attentions on the challenges of implementation.

By advancing incrementally, the system grew bigger and the technology grew more sophisticated. Once the ATMS had proven itself on 10 intersections, it expanded to 100 and then 200. Once the staff had mastered fixed signal control, they advanced to adaptive signal control. Later they moved to advanced technologies (for example, GPS systems). The traffic control center, originally located in a closet in the fire house, evolved into the electronic nerve center for the county.

Technological change required and promoted organizational change. MCDOT initially had to develop expertise in-house in order to accommodate the technology. That expertise, in turn, stimulated a host of new applications of computer technology in MCDOT. As outside organizations, such as police and transit, used the capabilities of the ATMS, they began adapting to take advantage of the system. As a result, the new technology stimulated new inter-organizational relationships and internal reorganizations.

Furthermore, as the system grew, it leveraged other "adjacent" systems to gain additional functionality at little extra cost. By linking video images into the cable television system, traffic information could be disseminated at little additional cost. By connecting to local television stations, live broadcasts on the evening news shows became possible. By connecting operators' consoles to the Internet, traffic information can be accessed by anyone on the Internet.

Thus the development process involved incremental change in technology and organization. Although the system, as it exists today, is technologically advanced, this achievement was realized through attention to organizational factors rather than just to technology.

The ATMS and the National ITS Program

Although the Montgomery County ATMS precedes the FHWA's national ITS program by 13 years, it shares many of the same goals. Relative to traffic, the system attempts to increase the capacity of the road network, reduce delays, and shorten response times to incidents. In respect to environmental goals, the ATMS aims to reduce fuel consumption, improve air quality, and improve mass transit operations. The system also seeks to improve safety. Finally, ATMS attempts to improve regional transportation integration and promote information sharing. These goals are shared by the national program and the county's system development efforts.

Montgomery County's experiences illustrate important differences from the FHWA program. Foremost, MCDOT is less technology-oriented than the FHWA. MCDOT began by implementing new, but not advanced, technologies. The staff invested their efforts in the implementation of existing technology, not in field testing new technology. The national program has, from its inception, emphasized technological rather than organizational innovation.

A second difference, related to the first, is that MCDOT took an incremental approach to development. The federal program began with a vision of a total system, whereas MCDOT developed a broad system through periodic adaptations. Although MCDOT did take steps to allow for future growth, such as using an open architecture, the ATMS in the late 1990s was not so much the realization of a goal as the product of a long process of learning and incremental change.

Essentially, MCDOT's system is up and running today, while most projects in the national program remain in early stages of planning and development. With nearly 20 years of work performed, the ATMS exemplifies what ITS can do in practice. As a result, MCDOT's ATMS allows for *ex post* evaluation – as an existing system it can be assessed based on real data – although such an evaluation has not yet been performed and remains difficult to perform. In contrast, most systems in the national program allow only for *ex ante* evaluation based on their vision, rather than on operational experience that is still limited.

EVALUATION ISSUES

The strength of a multi-criteria evaluation methodology like that adopted in this book is that it can incorporate many more benefits (and possibly costs) in its assessment than traditional cost-benefit methodologies. An evaluation of Montgomery County's ATMS with the multi-criteria methodology illustrates this. For example, in addition to traditional benefits for traffic capacity, the

ATMS has strong benefits in the non-traditional category of institutional outcomes. However, the multi-criteria approach does not solve all problems of evaluation. The same difficulties present in traditional evaluations and in isolating and measuring benefits also pose problems for the multi-criteria approach adopted here.

Summary of Overall Evaluations Based on Application of Checklist

Application of the checklist reveals four categories of benefits that are particularly salient. These are: "Congestion and delay", "Capacity and mobility", "Cost relative to other options", and "Institutional outcomes". Other categories also have some relevance, but considerably less than these three.

Congestion and delay
The MCDOT ATMS is intended to render the road network more efficient, thereby reducing traffic congestion and the associated driving delays. The four indicators in this category should all show positive benefits (peak hour VMT – up; vehicle-hours of delay – down; peak and non-peak travel times – down; average length of queues – down.)

Capacity and mobility
By rendering existing road usage more efficient, the ATMS should effectively increase capacity. Furthermore, by providing information to motorists in their homes mobility is enhanced. Ease of mobility increases as drivers can make informed decisions about when and how to travel based on information available in their homes or offices. System accessibility increases as the ATMS improves transit. Daily VMT is likely to increase as drivers take advantage of road capacity and travel convenience.

Cost relative to other options
Nearly all ITS technologies are much less expensive than traditional road-building approaches to realizing capacity increases. Montgomery County's ATMS is no exception. Construction of an ATMS is much less expensive than roads. The addition of ATMS communication capabilities costs $2000 per signalized intersection. Operations costs are also reduced, because the system greatly enhances operations. Operational performance and malfunctions can be detected electronically without a site visit. Maintenance has not proven unduly expensive. However, it must be noted that since the system is still expanding, MCDOT has an abundance of in-house technical resources. Once the system has been fully operational for some years, the in-house skill base could decline and maintenance could pose a greater challenge.

Institutional outcomes

This is an important category of benefits for the ATMS. The flexibility and adaptability of MCDOT has been greatly enhanced. As noted above, the ATMS provides both rapid notification of changes in traffic and a set of tools with which to respond to such changes. The system also exhibits numerous synergies of both a functional and institutional nature. MCDOT's ATMS has become integrated with cable television, the Internet, radio, and television. Institutional synergies have been created with police, fire, emergency, and the mass media. The synergistic benefits of the technology seem to be among its biggest benefits. Related to this, inter-governmental cooperation has been greatly enhanced by the systematic functionality of the ATMS. The ATMS has not given rise to ownership changes in a literal sense. However, over time MCDOT has acquired *de facto* "ownership" of much of the county's telecommunications policy and is now pioneering a county-wide information infrastructure. Conceived of as "jurisdiction," ownership has evolved as a result of the ATMS.

Public–private partnerships

Such partnerships have not been a prominent feature of the ATMS. Early in its development, MCDOT worked closely with the original system vendor. The system, however, must be seen as a model of private sector entrepreneurship. There have been some liability concerns with the system, insofar as the mere existence of traffic control capabilities provides an opportunity for lawsuits when accidents occur. However, there have been no significant legal actions to date.

Finally, privacy has not been an issue for the system, although it could become one. The ATMS collects no financial or registration data, and therefore is not linked to many of the most common privacy issues. However, it does employ a large number of surveillance cameras mounted on poles in public spaces. In order to educate the public, MCDOT staff have made efforts to explain the system through citizen group tours and outreach meetings. Nonetheless, the following letter to the local *Montgomery Journal* (November 14, 1991) expresses well the uneasiness that such a system could provoke: although the tone is tongue-in-cheek, the underlying concerns expressed are very real:

[Dear Sirs]:
  Recently, while walking across the Falls Road bridge over Interstate 270, I was checking the sky for UFOs and saw something peculiar. It was not a spacecraft, but a singular mysterious metallic object mounted high atop a telephone pole. The pole is west of the overpass. I would not really have noticed it, but there was nothing else on this pole except the object. No cables, nothing.
  I stared at it for a while, and then the strangest thing happened. It moved. It

moved! I swear it did.

What the heck is this thing? I think it must be a camera. If not, my guess is that it was set up by the National Security Agency to intercept highway cellular phone conversations. Am I right?

It seems likely that a system with 200 video cameras mounted alongside public roads could eventually provoke public concerns with privacy. This may have as much to do with the visibility of the cameras than with actual privacy violations that might occur.

These four categories of criteria are the most relevant for evaluation of the ATMS. The other categories in the multi-criteria methodology also have some bearing, but less than these four.

Safety
Improved system responsiveness is likely to have significant safety benefits. As delays are reduced, and incidents are cleared away more quickly, associated safety hazards (for example, deaths, injuries, and property damage) are also likely to decrease.

Financial and fiscal outcomes
The Montgomery County Office of Management and Budget thoroughly reviewed the ATMS development program before approving its budget. Financial analysts acknowledged the difficulty of quantifying the benefits of the system, but were convinced that overall financial and other returns would be substantial. The system does not generate new revenue for the county, and therefore cannot be said to have benefits in terms of immediate revenue enhancement or budgetary sufficiency. However, the benefits that it offers to other departments were thought to be significant. For example, police officers now spend less time directing traffic around incidents, and they are better informed when responding to traffic emergencies. In effect, there were intergovernmental transfers of benefits taking place, although they could not be easily measured. Relative to the private sector, the ATMS has not been a source of private profits, nor was this a goal of the ATMS. Some procurement contracts were issued to a local firm, however (Orbital Sciences Corporation in Germantown, Maryland).

Environmental outcomes
Since the ATMS operates at the county level and not the regional level, environmental issues have not figured prominently in MCDOT's planning. Air quality and clean air regulations have greater salience at the regional (greater Washington) level. Ambient emissions levels are probably lower, as congestion is reduced and traffic flows are smoother. Ground-level ozone may also

be reduced. Soil erosion, wetlands and habitats are unaffected by the system.

Economic outcomes

The relationship between improved transportation infrastructure and economic development is ambiguous. This ambiguity is even greater in the case of a technology that marginally enhances existing infrastructure. The ATMS probably has no measurable effect on economic outcomes (gross regional product, employment, per capita income, land values, and regional competitiveness), although some local economic benefits may have been realized by the award of procurement contracts to local suppliers.

Equity outcome

The ATMS is supported by general tax revenue, and its benefits go to all road users (including all residents in a suburban county as well as others). Therefore, the correlation between who pays and who benefits seems equitable. Since individual citizens do not purchase any in-vehicle systems, affordability is not an issue here. With the expansion plans covering the entire county, regional access disparity has been avoided. (Note that handicap access is not relevant to this technology.)

ASSESSMENT OF MAJOR EVALUATION DIMENSIONS

Although the multi-criteria methodology broadens the range of factors considered in evaluation, the application of the additional criteria, however, can pose problems: many costs and benefits can be measured or estimated, many others resist such measurement. The ability to measure benefits is unevenly distributed across categories of benefits.

Practical Measures

To date, MCDOT has not attempted a comprehensive evaluation of its ATMS. Nonetheless, it does employ a variety of measures as indicators of the performance of the system. These measures are:

- responsiveness;
- complaints of signal malfunctions;
- reduced time investigating complaints.

The system's greatest benefits derive from increased responsiveness. This evaluation is based on a transportation engineering heuristic: MCDOT staff estimate that one minute of delay in reacting to traffic congestion (usually an incident) corresponds to five minutes of traffic congestion. This heuristic allows a translation of a measurable quantity into a measure of congestion reduction. They conclude that the ATMS reduces reaction times, and therefore that it reduces congestion.

Another measure of effectiveness is the number of complaints about traffic signal operations. Over a six-year period complaints dropped from 3000 per year to 300 per year, which reflects improvements in the operation and maintenance of traffic signals.

Another related measure is the amount of time needed to investigate a complaint. Before the ATMS was installed, a technician might spend an entire afternoon investigating a complaint. With the ATMS, the functioning of signals and operations can be checked remotely from the Transportation Management Center. Airplane observations have also dramatically reduced response times and increased the number of sites that can be observed in a given time period. Ultimately, this saves staff time even while increasing MCDOT's responsiveness.

Most of these measures are related to just one category of evaluation – cost relative to other options. The other three of the four relevant categories prove more difficult to measure.

Methodological barriers arise when measuring congestion and delay and capacity and mobility: the problem is relating the functions of the ATMS to specific changes in transport. Isolating the cause-and-effect relationship is difficult. The incremental strategy used to develop the system caused benefits to be realized incrementally as well; as a result, beneficial changes in the transport system are indistinguishable from other changes taking place in the transport system. In 17 years of development, the system has probably improved many aspects of road transport. However, these effects cannot be isolated, because so much else has changed as well.

In theory, causality could be tested by a two-part experiment. First, all the measures above would be performed with the ATMS in operation. Then the system could be shut down, and the measures could be performed again (perhaps after a waiting period to allow traffic to "settle"). Presumably, differences in the measures would reflect the removal of the ATMS. This would be an ideal way to examine the counterfactual.

However, such an experiment cannot be conducted in practice, due to the adverse effect it would have on the county's traffic. With the ATMS down, public service would be compromised. Moreover, there are no regularly scheduled opportunities to observe traffic without the ATMS. Even when the ATMS is brought down for scheduled maintenance, a backup system keeps it operational. Nor does a system accident offer an opportunity for comparison: the backup system prevents accidental shut-downs quite effectively. And in the rare event of a system shut-down, traffic data collection ceases as a result, so that no measures exist of the altered traffic.

Measures of institutional outcomes are also elusive. There are clear benefits from receiving an aerial video download so that police officers at an emergency site can make better decisions. Greater inter-agency coordination, in general, also is useful. However, such institutional effects cannot be meaningfully quantified.

Thus, some difficulties of measurement remain with the evaluation methodology adopted for this book. However, by directing attention to additional categories of costs and benefits, this method illuminates important categories of benefits that might otherwise not enter into an evaluation.

Practical Measures for Alternative Evaluation

The history of the ATMS suggests an additional type of measurement. In addition to evaluating the system as it exists at a moment in time, it can be evaluated as it grows over time. This requires quite different criteria and indicators.

The history of MCDOT gives insights into how ITS technology is successfully deployed, showing how a long-term development path begins and how it evolves. This implies insights into both the mechanisms by which system development shapes future activities and the types of long-term benefits that can result from ATMS activity.

MCDOT and its ATMS did not have to develop as they did. Strict relationships of cause and effect were not at work, and the series of decisions were neither necessary nor predetermined. However, in analyzing development – even if one cannot work with cause and effect – insights can be acquired. Conditions that made development possible, if not necessary, can be identified. In comparison, other systems can be evaluated to determine if they have successfully acquired the enabling conditions that proved important in Montgomery County.

The first enabling condition consisted of an external source of technical initiative. MCDOT received a stimulus from the mall developer. Undoubtedly, MCDOT staff were receptive to the suggestion for new technology, but the larger plan only emerged once an outsider had proposed and funded a system. The availability of external expertise and funding may have enabled this first step.

Another important factor in development was the ability of the MCDOT staff to learn. They encountered numerous difficulties, but overcame them. In doing this, their close relationship with the system vendor proved important: when they encountered difficulties, they could get assistance. This collaborative relationship may have facilitated learning.

The development of collaborative relationships with other agencies seems to have capacitated later initiatives. Each generation of technology promoted a greater degree of collaboration, and this in turn enabled a later generation of technology. The formation of relationships is an important enabling condition.

This enabling condition can be translated into operation measures for evaluation which can be used to assess whether a transportation organization has performed a "good" field test. The following list identifies some evaluative criteria and practical indicators for measuring system development.

- *Has expertise taken hold in an organization?* Field tests should create organizational expertise so that further activities can be undertaken by staff. *Measures:* How much technical activity was performed by internal staff vs. external consultants? Has the budget of ITS technical staff increased? Have staff received additional education? Do favorable career tracks exist for technical staff?
- *Has commitment formed in the organization?* If a field test is to be followed by additional organizational initiative, then top-level policy makers must support ITS. *Measures:* How much funding did the organization contribute to the field test? Was there any history of ITS activity prior to the existence of federal funds for field tests? Has outreach been performed to local legislators who make funding decisions?
- *Have relationships formed with external technical experts?* If the organization is to continue with development, it will continue to be dependent on outside assistance for some time. To survive this period of dependence, it may need reliable external technical assistance. *Measures:* Longevity of internal and external staff on a project (that is, lack of turnover) and frequency of communications (telephone, e-mail).
- *Have relationships formed with other public agencies?* Many benefits of ITS may not be realized without collaboration with other agencies. Relationships with agencies enable continued development. *Measures:* Increase in interagency committees. Jointly funded activities and evidence of mutual adjustment, negotiation, and compromise.

These are indices that can be used to evaluate whether a field test is likely to be effective in initiating a process of ITS development. Although they cannot provide definitive data about the future, they may indicate whether a field test is more or is less likely to lead to continued development. Together with the multi-criteria evaluation method, this method may help analysts decide whether a system and a development program show promise.

CONCLUSIONS

The multi-criteria method for ITS evaluation adopted here broadens the project dimensions subject to evaluation, thereby capturing more of the total costs and benefits of ITS. This was evident in its application to the Montgomery County ATMS, in which four categories of benefits were highly significant. Although three categories dealt with fairly traditional topics (congestion, capacity, and relative costs), the category of "Institutional outcomes" identified many non-traditional costs and benefits. Traditional methods would not have included this criterion in the evaluation.

Some problems remain in translating these criteria into practical measures, however. Although new inter-organizational coordination capabilities have been created, measures of cost and benefit for these remain elusive. Still, some distinct measures do provide indicators of the effects of the ATMS. MCDOT used a variety of measures to evaluate the impacts of the ATMS. These included: reaction times to traffic incidents, number of citizen complaints about signal malfunctions, and time needed to investigate complaints. All these showed positive benefits from the ATMS.

Some additional evaluation criteria have been proposed for measuring the evolution of the system over time. These are more qualitative in nature, indicating whether an organization is more or less likely to succeed in ITS development. These measure such factors as expertise, commitment, and relationship-formation.

In conclusion, the evidence of the Montgomery County ATMS indicates that the application of the multi-criteria approach allows for a more complete evaluation than traditional measures.

APPENDIX: MULTI-ATTRIBUTE CHECKLIST WITH MOST IMPORTANT EFFECTS INDICATED

Category	Goal	Indicator	Effect	Comment
Congestion	—	Peak-hour VMT	Up	—
and delay	—	Vehicle-hours of delay	Down	—
	—	Volume/capacity ratio	Better	—
	—	Peak/non-peak travel times	Down	—

Category	Goal	Indicator	Effect	Comment
	—	Average speed (for example PMT/P-Hr)	Up	—
	—	Average length of queue	Down	—
Capacity and mobility	—	Ease of personal mobility	Better	—
	—	System accessibility	Better	—
	—	Network or connectivity implications	—	—
	—	Passenger ridership	—	—
	—	Daily VMT	Up	—
Safety and accidents	—	Deaths	—	—
	—	Injuries	—	—
	—	Property damage	—	—
Cost relative to other options	—	Construction	—	—
	—	Operations	—	—
	—	Maintenance	—	—
Financial and fiscal outcomes	—	Revenue enhancement	—	—
	—	Budgetary sufficiency	—	—
	—	Inter-governmental transfers	—	—
	—	Private profits	—	—
Environmental outcomes	—	Total emissions	—	—
	—	Ground level ozone	—	—
	—	Energy consumption	—	—
	—	Hazardous materials impact	—	—
	—	Soil erosion	—	—
	—	Wetlands and habitat	—	—
Economic outcomes	—	Gross regional product	—	—
	—	Employment	—	—
	—	Per capita income	—	—

Category	Goal	Indicator	Effect	Comment
	—	Land values	—	—
	—	Regional competitiveness	—	—
Equity outcomes	—	Affordability	—	—
	—	Regional access disparity	—	—
	—	Handicap access	—	—
Institutional Outcomes	—	Flexibility, adapts	Better	—
	—	Regional significance		—
	—	Synergy	Better	—
	—	Intergovernmental friction/cooperation	Better	—
	—	Community involvement and preferences	—	—
	—	Ownership changes	Better	—
	—	Public–private partners	—	—
	—	Liability	—	—
	—	Privacy	Worse	—
Spatial outcomes	—	Land use implications	—	—
	—	Land values	—	—
	—	Residential proximity issues	—	—
	—	Infrastructure design or geometry impact	—	
Demonstration outcomes	—	Transferability	—	—
	—	Value as a model	—	—
	—	Suitability	—	—

NOTES

1. The web address is http://www.dpwt.com/traffpkgdiv/camera.html.
2. See Neeman and Huang, 1993, and Lukasik, 1997, for a discussion of information management and integrated traffic management systems.
3. Op. cit.

4. Traffic signalization is at the heart of ATMS systems. References to some of the research addressing signalization management appear in Chin *et al.* (1999); Su and Chang (1996); Taylor and Wolshon (1999).
5. See note 4.

REFERENCES

Adeyinka, Olumide (1997), "Undefeated by traffic: strategic ATMS for Montgomery County", *Traffic Technology International* August/September, 85–8.

Awadallah, F. and Habesch, N. (1998), "Evaluation of video image processing for traffic management systems", *Transportation Quarterly* **52**(3), 79–90.

Berka, S, and Lall, B.K. (1998), "New Perspectives for ATMS: advanced technologies in traffic detection", *Journal of Transportation Engineering* **124**(1), 9–15.

Chin, Daniel C., Smith, Richard H. and Spall, James C. (1999), "Evaluation of system-wide traffic signal control using stochastic optimization and neural networks", in *Proceedings of the 18th American Control Conference*, Evanston, IL: American Automatic Control Council, pp. 2188–94.

Donaldson, G.S. (1992), "Traffic management in Montgomery County, Maryland for the 90s and beyond", *Compendium of Technical Papers*, Washington, DC: Institute of Transportation Engineers, p. vi.

Florida Department of Transportation (1995), *Statewide Communications Protocol Study, Final Report*, Report no. state job 99700-3516-119, WPI 0510671, March.

Havinoviski, G.N., Nguyen, V.T. and Sutaria, T.C. (1996), "Building a platform for ITS: Santa Ana's Advanced Traffic Management System", paper presented at *Intelligent Transportation: Realizing the Benefits*, Proceedings of the 1996 Annual Meeting of ITS America, sponsored by ITS America, Houston, TX, May.

Hill, T., King, M., Lovelace, J.L., Roberts, E.G., Schauer, M.D. and Shyne, S.S. (1996), "Traffic flow visualization and control (TFVC) improves traffic data acquisition and incident detection", paper presented at *Intelligent Transportation: Realizing the Benefits*, Proceedings of the 1996 Annual Meeting of ITS America, sponsored by ITS America, Houston, TX, May.

Hyman, W.A. and Miller, N.P. (1996), "Monopoly franchises vs. competitive models for deployment of advanced traffic management systems and advanced traveler information systems", paper presented at *Intelligent Transportation: Realizing the Future*, abstracts of the Third World Congress on Intelligent Transport Systems, sponsored by ITS America, Orlando, FL, October.

Jacobson, L. and Sumner, R.L. (1995), "Multijurisdictional information system for traffic", presented at *Intelligent Transportation: Serving the User Through Deployment*, Proceedings of the 1995 Annual Meeting of America, sponsored by ITS America.

Kaiser, P.D. (1999), "Management system for the Las Vegas Valley", *ITE Journal* **69**(4), 26–9.

Kerr, J., Khosravi, F.E. and Simone, C. (1997), "CALTRANS District 12 Advanced Transportation Management System: solving Orange County's transportation problems", presented at *Merging the Transportation and Communications Revolutions*, Abstracts for ITS America Seventh Annual Meeting and Exposition, sponsored by ITS America, Washington, DC, June.

Knopp, M.C. and Lynch, S. (1996), "Impediments to ATMS deployment in the US: A perspective from the field", in *Intelligent Transportation: Realizing the Future*, abstracts of the Third World Congress on Intelligent Transport Systems, sponsored by ITS America, Orlando, FL, Oct.

Lukasik, D.A. (1997), "The CALTRANS District 7 Advanced Traffic Management System communications design", paper presented at *Merging the Transportation and Communications Revolutions*, abstracts for ITS America Seventh Annual Meeting and Exposition, sponsored by ITS America. Washington, DC, June.

Manners, M.A. (1994), "An innovative trend in traffic management and incident detection", *IMSA Journal* **38**(2), 44–6.

Mitchell, D. (1998), "Transport Advanced Transportation Management System program", *ITE Journal* **68**(9), 48.

Mohaddes, A. (1995), "Consensus building process in the development of ATMS/ATIS", paper presented at *Steps Forward*, Intelligent Transport Systems World Congress, Yokohama, Japan, Nov.

Neeman, Barbara and Huang, Greta P.Y. (1993), "Information requirements for an integrated transit/traffic management and traveler information system", *IVHS Journal* **1**(2), 167–80.

Pittenger, Jerry (1997), "Atlanta's ITS showcase: the project that worked", *ITS Quarterly* **5**(1), 41–4.

Su, Chih-Chiang and Chang, Gang-Len (1996), "An integrated model for adaptive intersection signal control", *ITS Journal* **3**(3), 224–48.

Sussman, Joseph M. (1999), "Regional ITS architecture consistency: what should it mean?", *ITS Quarterly* **7**(3), 3–5.

Taylor, W.C. and Wolshon, B. (1999), "Analysis of intersection delay under real-time adaptive signal control", *Transportation Research. Part C: Emerging Technologies* **7**(1), 53–72.

Turnbull, K.F. (1997), "Workshop on integrating transit with advanced traffic management systems: workshop proceedings", May, Report no. interim report TTI/ITS RCE-97/05.

Watje, John M., Symes, Denis and Ow, Robert S. (1994), "Vehicle location technologies in automatic vehicle monitoring and management systems", *IVHS Journal* **1**(3), 295–304.

PART II

Applications of Alternative Methodologies to
ITS

7. Smart Flexible Integrated Real-time Enhanced System (SaFIRES)

Laurie Schintler

INTRODUCTION

In September 1995, the Potomac Rappahannock Transportation Commission (PRTC) in Northern Virginia began implementing the Smart Flexible Integrated Real-time Enhancement System (SaFIRES), a project designed to enhance the performance of an already existing flexible-route transit system. SaFIRES is a real-time request, scheduling and dispatching system that uses such ITS technologies as: a Global Positioning System (GPS)-based automated vehicle location (AVL); real-time scheduling software; GIS mapping; digital communication via mobile data terminals; and integrated computerized dispatching software.

This study evaluates SaFIRES in the context of national ITS and project-specific goals and objectives. At the time of this study (1995–6), the operational test was not due to be completed for two years and, therefore, the evaluation was primarily *ex ante* – although preliminary impacts of the project were included in the evaluation. A multi-criteria checklist analysis was used to evaluate the project along a number of dimensions: congestion and delay; capacity and mobility; safety; operating and capital costs; financial and fiscal outcomes; the environment; the economy; equity; and institutional arrangements and structures. The study also demonstrates how a goals-achievement method can be used in conjunction with the checklist to determine whether or not a project of this nature should be implemented.

HISTORICAL CONTEXT AND PROJECT OVERVIEW

The study area, comprised primarily of Prince William County, located about 25 miles south-west of the District of Columbia, is a fast-growing, automobile-oriented, suburban residential community. Residential density is relatively sparse at about three persons per acre. Commercial activities are limited mainly to retail and service and the employed commute to areas outside the county.[1] Prior to 1990, this area had no public transit system, except for taxicabs. Service began in December 1994 with three "flag-stop" feeder routes to two commuter rail stations. In April 1995, flexroutes on three major commuting corridors were also implemented, and subsequently expanded to six corridors and five feeder routes. The fare structure for each of these services is presented in Table 7.1.

Table 7.1   Current fare structure

Upon boarding service	Fare required ($)
Feeder or route deviation with no intent to transfer	0.075
Feeder or route deviation with an intent to transfer	0.075 plus $0.25
Feeder or route deviation with token or cash for the commute/ride fare	0.25
Feeder or route deviation with valid VRE ticket	0.25

In September 1995, the PRTC began implementing the SaFIRES enhanced transit system in accordance with a two-year, three-phase schedule. The three phases are:

1. planning and system design;
2. implementation of a non-ITS-enhanced flexible-route transit system;
3. implementation and testing of the ITS-enhanced system.

Both the first and second phases were completed on schedule, and the third phase has been initiated.

Institutional Setting

Prior to September 1995, the Prince William County transit system was managed by the Potomac and Rappahannock Transportation Commission (PRTC), a commission funded by the public sector, and operated under contract by the Mayflower/Laidlaw, Inc. The Northern Virginia Planning District Commission (NVPDC) played a supportive role, providing local GIS plotted market research data used in determining OmniLink routes. Several public and private organizations were involved with the operational testing of SaFIRES until (phase 3) implementation in September of 1996. The public institutions involved in the project included: the Federal Highway Administration (FHWA); Federal Transit Administration (FTA); Northern Virginia Planning District Commission (NVPDC); and the Virginia Department of Rail and Public Transportation (VDRPT). Private partners include GMSI, Inc.; Castle Rock Consultants; SG Associates, Inc.; Tidewater Consultants, Inc. (TCI); and UMA Engineering, Ltd.

PRTC is the overall project manager funding 67 percent of the total cost (approximately $2.9 million), and FHWA and FTA are serving both as technical advisors and financiers, providing 28 percent ($1.2 million) of the project's cost. Private sector parties paid the balance of total expenses, contributing almost $200 000 to the project. Mayflower/Laidlaw has continued to operate the mass transit system under contract, and NVPDC also continued to provide GIS maps of the local area. However, NVPDC began charging a fee for monitoring, documenting, and relaying any findings to OmniRide after September 1996.

VDRPT served as the funding conduit and also provided project oversight services. GMSI, Inc. provided software and hardware that was necessary for the installation and application of mobile data terminals, GPS-related hardware, and card and odometer readers. GMSI worked with UMA to assure proper and efficient system integration; UMA also developed a computer-based scheduling and dispatching computer system for the project. SG Associates, Inc. assisted PRTC with project management, serving as a consultant on issues related to transit configuration and operations applications, and conducted a peer review of the project. TCI provided overall project oversight consisting of: supervision of the testing and integration of system components; general technical support; and the identification of problems and their root causes.

Evaluation of SaFIRES

SaFIRES was one of many projects selected by the Federal Highway Administration to be part of the National APTS ITS program. The goals and objectives of the national APTS program are as follows:

Objective 1: To enhance the quality of on-street services to customers.
   Sub-objectives:
   a.   improve timeliness and availability of customer information;
   b.   increase convenience of fare payments;
   c.   improve safety and security;
   d.   reduce passenger travel time;
   e.   enhance opportunities for customer feedback.

Objective 2: To improve system productivity and job satisfaction.
   Sub-objectives:
   a.   reduce transit system costs;

b. improve schedule adherence and incident response;
c. increase usefulness of data for planning and scheduling;
d. enhance response to vehicle and facility failures;
e. improve information management systems and management practices;
f. reduce worker stress and increase job satisfaction.

Objective 3: To enhance the contribution of transit to overall community goals.
Sub-objectives:
a. provide discount fares to special groups;
b. improve communication with users having disabilities;
c. enhance the mobility of users with disabilities;
d. increase the extent and effectiveness of TDM programs;
e. enhance HOV systems by reducing SOV use;
f. assist in achieving air quality and energy goals and mandates.

Goals and objectives specific to the Prince William County project

1. Improved on-street and in-office efficiencies.
2. Enhanced vehicle tracking and communication between bus drivers, dispatchers, and other service providers.
3. Improved access to transit.
4. Simplified reporting and tracking of human service ridership and agency charges.
5. Promote synergy between human service agencies.
6. A reduction in the advance reservation time for flex-route services from between 24 and 48 hours to one hour. ("SaFIRES", 1995)

SUMMARY OF EVALUATION OUTCOMES

Based on these goals and objectives, SaFIRES was evaluated using a checklist approach. Only the most significant outcome impacts are discussed here; however, the reader may consult the Appendix for a summary of all outcomes.

Congestion and delay   ITS-enhanced bus services have a strong potential to reduce congestion and delay (Kuhl, 1995). This could result if the improvements in service offered by the ITS enhancements are great enough to encourage substantial shifts in mode choice, in particular, from drive alone to transit (that is, bus). These improvements would consist of: (a) a reduction in paratransit service request time; (b) lower operating costs; (c) decreased peak travel times.

Paratransit service request time   The Global Positioning System (GPS)-based automated vehicle location (AVL), real-time scheduling software, digital communication via mobile data terminals, and integrated computerized dispatching software could reduce the amount of time required (in advance) to request paratransit service. At time of implementation, reservations had to be made 24–48 hours in advance. With ITS enhancements, this could be reduced to one hour ("SaFIRES", 1995).

Operating costs   ITS enhancements also have the potential to reduce operating costs, which would allow for service improvements. Reductions in operating costs may also prevent the need for raising transit fares.

Peak travel times   Peak and non-peak hour travel times are also likely to decrease: as a result of more efficient routing, regular patrons of the system may experience shorter travel times. New users of the service, specifically individuals who have previously driven alone, may encounter reductions in travel time. Any resulting reductions in congestion may also decrease travel times.

Ridership   To evaluate SaFIRES, in terms of its effect on ridership, and indirectly on congestion and delay, one could compare conditions before and after implementation. One challenge of comparison would be to determine whether future increases in ridership, or modal shifts, were caused by the ITS enhancements themselves or by natural increases in population and employment. Within the last few years, Prince William County has grown significantly in terms of population and employment, and is expected to do so in the future, therefore the growth factor must be weighted in evaluation (Metropolitan Washington Council of Governments, 1998). Ease of personal mobility and system accessibility may be improved for all of the reasons stated, and if improvements in operating efficiency lead to better transit services to METRO, then interregional mobility will also be enhanced.

Environmental impact   The potential environmental effects of SaFIRES could include: lower ambient emissions levels, as a result of improvements in traffic flow or shifts in mode choice (that is, from drive alone to mass transit); reduced soil erosion and damage to wetlands and habitat, to the extent that it prevents the need for further lane or highway expansion.

COSTS OF IMPLEMENTATION AND OPERATION

There are three different types of technologies or approaches which may be used in Automated Vehicle Location (AVL): the signpost approach, Loran-C, and GPS, with cost and efficiency tradeoffs associated with each use:

The signpost approach   This system, using a series of sign mounted utility posts approximately 11–16 feet above the street detects the presence of a vehicle within a 300–600 foot radius. However, there are a number of high maintenance costs associated with this approach; batteries in the signpost devices must be replaced frequently, and the signposts themselves must be tested and replaced regularly. Although a less costly approach, the signpost, given its fixed location, is not very practical for tracking demand-responsive transit vehicles.

Loran-C   Loran-C is a land-based radio navigation system which uses low-frequency radio waves to provide signal coverage on land. The Loran receiver collects information on radio waves; the waves are then used to determine the location of a moving object. The system, which can provide location accuracy within 500 meters, has proven to be an acceptable method of tracking buses in non-congested metropolitan areas.

Global Positioning System   The GPS, which is used in the SaFIRES project to locate vehicles, is the most flexible and accurate of the three approaches.

Defining Vehicle Intelligence

There are generally two ways to define vehicle intelligence in an ITS-enhanced transit system, each of which uses a different set of technologies and mode of operation. The first approach uses a "smart bus". The second approach, adopted by SaFIRES, uses a "dumb bus". In comparison of costs between the two systems, the "smart bus" has a higher implementation cost but possibly a lower operating cost (Smith, 1991):

Smart bus   A "smart bus" is a bus equipped with an on-board computer, capable of data collection and analysis. The benefits may include: a reduction in data transmission; an increase in the number of buses which can be covered by any given base station; fewer radio frequencies; a reduction in data storage at the base station; and immediate feedback on operation problems.

Dumb bus   A "dumb bus" acts as a probe, relaying all information to a central control center for analysis. A "dumb bus" may provide reductions in the cost of on-board computers and decreases in the need to load and reload bus schedules. Furthermore, through the transmission of data at regular intervals a full travel history can be maintained with this approach.

Revenue Enhancement and Budget Sufficiency

The fiscal and financial impacts of SaFIRES should be fairly significant for both public and private sector agencies. Although there is an initial fixed cost associated with implementing this ITS enhanced system, long-run gains in operating efficiencies and reductions in operating costs should offset these costs. Operating costs may be decreased in one of two ways: (a) by maintaining the same service, in terms of area and frequency, with a reduced number of buses; (b) by using the same number of buses, operating in a larger coverage area, to increase ridership and revenue (Smith, 1991). The extent to which operating costs can be reduced depends in part on the accuracy of the GPS; however, in areas with significantly tall buildings or tree canopies, location accuracy becomes less reliable. Obviously, for Prince William County, the first factor is not an issue, at least not yet; however, the second may be, as the county is still heavily forested.

Gains in operating efficiency should be accompanied by an increase in ridership, and consequently passenger revenues. More individuals will use the new mass transit system due to improvements in comfort and convenience and an expanded market area. These combined factors should decrease operating and overhead costs, thereby reducing the financial contributions required of Prince William County, the City of Manassas, and the City of Manassas Park. Substantial improvements in budgetary sufficiency should result. This may minimize the need to increase fares or reduce service levels which would of course benefit users of the system. The Toronto Transit Commission (TTC), which recently implemented an AVL and Computer-Aided Dispatch system similar to SaFIRES, has already seen increased use of its fleet (5–25 percent), multi-million dollar operating cost reductions, and significant passenger-revenue increases.

There are a number of quantitative models that can be used to evaluate the performance of paratransit, and with some modification could also be used to assess the performance of ITS-enhanced systems (Chang *et al.*, 1991). Also, there are several that can be used to quantify transit performance along a variety of dimensions, as described in Table 7.2.

Safety

With ITS enhancements, there are certain to be improvements in safety for bus drivers, passengers, and possibly even other motorists. The GPS will fa-

Table 7.2   Measures of transit performance

Performance dimension	Measure
Cost efficiency	Revenue vehicle hour per operating expense
	Total vehicle miles per operating expense
Service utilization	Unlinked passenger trips per revenue vehicle hour
	Unlinked passenger trips per revenue mile
Revenue generation	Corrected operating revenue per operating expense
	Operating revenue per operating subsidy
Labor efficiency	Total vehicle hours per total employees
	Revenue vehicle hours per operating employee
Vehicle efficiency	Total vehicle miles per peak vehicle
	Total vehicle hours per peak vehicle

Source: Parts of this table are from Fielding (1987)

cilitate real-time identification of incidents and major delays as they happen, which will enable emergency and service vehicles to be sent to the accident scene at a faster rate. Faster clean-up at the scene will minimize the chance of secondary accidents. In addition, emergency vehicles can be more efficiently dispatched to the precise location of the bus, if there is a mechanical problem with the bus itself or a medical emergency involving a passenger. Finally, with an emergency switch on board, bus drivers have the option of notifying the control center of any criminal activities or hijacking, allowing security officers to be dispatched immediately to the correct location. Located at the lower left side of the driver, this switch can be easily accessed by the driver in an emergency situation, but cannot be seen or heard by passengers. To measure safety improvements, deaths and injuries before and after ITS enhancements could be compared, although this would not be easy. A better way of gauging success in terms of safety may be to document experiences of drivers or passengers.

Equity

A more efficient, convenient, and comfortable transit service will improve mobility for economically and physically disadvantaged individuals. One goal of this operational test is to use GPS in monitoring the location of handicapped accessible vehicles in the network, and then use the information to dispatch the most appropriate vehicle – in terms of its on-board facilities[2] – to specific locations. However, equity outcomes are difficult to measure, particularly in this ITS application. Perhaps the best way to assess possible equity gains is to interview human service clients using the service. They could be questioned on issues, such as whether or not they believe that they had better access to mass transit than previously. In terms of economically disadvantaged individuals, improvements in operating efficiency may prevent fares from increasing, therefore maintaining the affordability of this transit service.

Synergy, Inter-governmental Cooperation, and Public–Private Partnerships

SaFIRES should simplify reporting and tracking of human service ridership and agency charges (Marx, 1995), therefore improving the operating efficiency of these agencies. Consolidation of human service transportation services, which is part of the operational test, may result in economies of agglomeration, stimulating additional economic benefits.

While SaFIRES is in operation, PRTC will be in contact with human service agencies. Inter-agency cooperation has been encouraged (for example, GMSI and UMA, two private firms, work together to assure proper and efficient integration of the technologies, and SG Associates, Inc. assists PRTC in managing the project). Many of these organizations are assuming managerial or supervisory roles, and each has an agenda and set of standards by which to gauge project success. It has been noted that the proportion of funds allocated to each agency is not commensurate with their responsibilities. If this mismatch truly exists, there could be friction, which would be a significant barrier to successful operation of the test. The extent of inter-agency cooperation may be measured by the number of committees or jointly funded activities that are formed during the test. One should also look for evidence of mutual adjustment, negotiation, and compromise (Klein, 1995).

Any increase in ridership by human service clients will increase inter-governmental transfers, as human service clients are not charged fares – on the contrary, their sponsoring agency is billed on a cost per trip basis (Marx, 1995). Private companies should also profit, particularly GMSI, Inc., a company that provided mobile data terminals, GPS-related hardware, card readers, and odometer readers, and UMA Engineering that developed the computer-based dispatching and scheduling software. The success of the operational test could stimulate additional demand for these products on the market.

Institutional Outcomes: Flexibility and Adaptation

SaFIRES is a highly flexible system, which in the future could be easily changed or combined with other ITS technologies. The Mobile Data Terminal, or MDT

4023, which was developed by GMSI, Inc., has an optional magnetic swipe credit card reader that could be used at a later date to bill transit users. This technology could also be used to read "smart cards", cards that could be used to pay for METRO parking.

Liability and Privacy

This technology raises some privacy issues. According to a series of court opinions, the right to privacy includes three interests: autonomy, intrusion, and informational privacy. Relating to intrusion, people are generally interested in being free from surveillance, specifically in circumstances where there is a reasonable expectation of privacy. Maintaining anonymity is a key aspect of this interest. Bus drivers may feel that constant monitoring of their position in the network is a violation of their privacy and symbolizes a lack of trust in their professionalism. One could argue that, despite the lack of anonymity, surveillance is in the interest of the driver, particularly for security and safety reasons. There may also be an issue of informational privacy, which concerns the agency that controls the collection, quality, use and dissemination of personal information (Belair *et al.*, 1993). Although magnetic swipe or "smart" cards were not be used in the initial stages of SaFIRES, there was an expectation that they would be used in the near future, with the objective of facilitating the flow of information pertaining to human resource riders.

Identification of Major Evaluation Issues

By applying the checklist method, it is apparent that with the type of project being evaluated here, there are certain outcomes that are likely to be more significant than others. The checklist helps significantly in the identification of the most important outcome dimensions. The pertinent outcomes include: mobility and accessibility – especially to economically and physically disadvantaged individuals; implementation and operating costs; safety; budgetary sufficiency; and interagency coordination and synergy. However, the checklist as a tool to support evaluation is of limited value when more qualitative precision is desired.

EVALUATING THE OUTCOMES

There are many techniques for evaluating and prioritizing alternative projects, given some set of project-specific outcomes. These include but are not limited to financial appraisal, cost-benefit analysis, cost-effectiveness analysis, planning balance sheet, and goals-achievement matrix. Descriptions of each of these techniques and assessment of their usefulness for evaluating ITS projects like SaFIRES appear below.

Financial Appraisal

The financial appraisal approach is based on profit maximization and cost minimization, where for each project an internal rate of return (IRR) is typically computed. The IRR associated with a project is the interest rate that balances present and future cash flows. This method ignores the social, environmental, and institutional impacts of a project, and is not well-suited for the evaluation of ITS projects.

Cost-effectiveness Approach

The cost-effectiveness approach assesses how each project contributes to the attainment of some set of goals and objectives. Associated with each objective is a measure of effectiveness (such as reduction in travel time). A cost-effectiveness ratio is a quantitative measure of how well a particular objective is met per dollar expenditure (Dickey, 1983). One limitation of this approach is that it does not provide a clear indication on the relative value of projects.

Cost-benefit Analysis

Cost-benefit analysis is a common technique for evaluating and prioritizing alternative projects. In an *ex ante* evaluation of multiple projects, this method entails: forecasting a stream of net benefits; discounting these net benefits; and summing over them to arrive at a "net present value (NPV)". A common decision rule is to select the project with the highest NPV, assuming this value is non-negative. Incremental methods, however, may be more appropriate when the projects being evaluated are mutually exclusive.

There are some difficulties associated with using cost-benefit analysis to evaluate Intelligent Transportation Systems projects. First, any costs and benefits used in the analysis must be in terms of some equivalent unit (for example, dollars). Many of the impacts related to ITS, such as improvements in inter-agency synergy, cannot easily be monetized. Second, it is difficult to determine by how much consumer surplus would increase with the implementation of ITS or, for example, an ITS-enhanced flexible-route transit system. To do this, the analyst requires an estimate of the price individuals are willing to pay for the service. In the case of transit fare – which is the price paid by individuals to use the service – this would not be a true reflection of demand. Nor is it an accurate indication of the costs of providing the service. For most transit systems, a large portion of this cost is subsidized. There are also distributional effects present, which cannot easily be incorporated into the cost-benefit analysis (for example, improvements in mobility for the economically and physically disadvantaged population at the expense of others in the community). Additionally, there are many benefits that cannot easily be quantified. These include: improvements in transit comfort and convenience; better interagency coordination; and increases in safety for passengers. Given the substantial fixed costs associated with this type of project (SaFIRES), if one looks solely at the quantifiable benefits, a cost-benefit analysis is likely to recommend staying with the status quo.

Planning Balance Sheet and Goals-achievement Method

The planning balance sheet, or community impact appraisal approach, does not require that impacts be measured in terms of a monetary equivalent. The method distinguishes social groups affected by costs and benefits and, therefore, clearly evaluates the distributional impacts. A goals-achievement approach is similar to the planning balance sheet, although it differs in that it completely avoids monetary values, focuses on the extent to which a project attains goals or objectives, and can weight the goals based on how the decision maker(s) would like to prioritize them (Dickey, 1983). This method seems appropriate for evaluating Intelligent Transportation Systems projects, particularly because the impacts do not have to be stated in terms of some equivalent unit; impacts do not have to be quantitative, that is, they may be subjective or even probabilistic. It should be noted, though, that the project can still incorporate the stream of benefits and costs may be included as impacts, taking into account the timing of these factors.

EVALUATING AN ITS-ENHANCED TRANSIT SYSTEM USING A GOALS-ACHIEVEMENT METHOD

This section demonstrates the use of a goals-achievement method, assuming there is some agency (such as PRTC) that can decide between two alternatives:

1. do nothing (operate existing flexible-route transit system in a manual dispatch mode, that is, without ITS technologies);
2. enhance existing flexible-route transit system with ITS technologies.

The first step in performing this type of evaluation is to obtain measures and estimates for each of the goals. The goals and criteria selected for this example correspond to those in column 2 of the checklist in the Appendix; the

Table 7.3   Hypothetical impacts of the two alternatives

Criterion	Alternative	
	I	II
1. Response time (minutes)	30	10
2. Cost/revenue-mile ($)	2.15	2.00
3. Cost/passenger-trip ($)	4.87	4.00
4. On-time performance – propensity for bus to be on time at each stop (%)	70	95
5. In-advance reservation time required (hours)	24	2
6. Initial costs (ITS equipment) ($ thousand)	100	2000
7. Continuing costs – maintenance of equipment ($ thousand/year)	20	200
8. Organizational coordination outside of transit agency (synergy) (on scale of 1 to 10)	6	8
9. Number of trips by disabled per month	200	250

measures or impacts are estimates based on the Prince William County Transit System. These outcomes are illustrated in Table 7.3.

Next, the goals-achievement effectiveness measures must be normalized and weighted. To derive normalized measures of goals achievement, the best level of each factor must be assigned a value of 100, while the scores on the other project(s) must be assigned a value that is a portion of that maximum. To account for the relative importance of objectives, each normalized value must be multiplied by a weight, indicating its importance. The results of this step for the hypothetical example presented here are illustrated in Table 7.4. Weighting scheme A assumes that each objective is equally important to the decision maker, while scheme B assumes that only the costs and benefits which are most likely to be included in a cost-benefit analysis are most important.

The rankings of the alternatives I and II by weighted scores is presented in Table 7.5. It can be seen that when only the costs and monetary benefits are strongly considered, this approach indicates that the transit system should continue to operate in a manual dispatch mode, yet when other factors, such as increased mobility for disabled individuals (criterion 10) are included, an ITS-enhanced system is more favorable.

Table 7.4   Normalized goals-achievement effectiveness measures for two
weighting schemes

Criterion	Alternative		Weighting schemes	
	I	II	A	B
1	33	100	10	2.5
2	88	100	10	15
3	80	100	10	15
4	82	100	10	15
5	77	100	10	15
6	8	100	10	2.5
7	100	5	10	15
8	100	10	10	15
9	75	100	10	2.5
10	80	100	10	2.5

Table 7.5   Comparison of alternatives under different weighting schemes

Weighting schemes	Alternative	
	I	II
A	7230	8150
B	8395	7000

CONCLUSIONS

The purpose of this evaluation is not to recommend that SaFIRES continue to
operate, or that such a system should be implemented in another locale. Rather,
it is to demonstrate what impacts are important to consider when evaluating
an ITS-enhanced transit system, and how such systems can be evaluated. There
are multiple challenges in evaluating such systems. Some of the outcomes,
such as impact on privacy, changes in inter-agency synergy, and so on, cannot
easily be converted to quantitative values or monetized. Therefore, techniques
like financial appraisal and cost-benefit analysis, which have these requirements, may not be the most appropriate for evaluating Intelligent Transportation Systems. Further, cost-benefit analysis requires as input an estimate of
how consumer surplus will change given implementation of the project. Because Intelligent Transportation Systems are still relatively new, it is difficult
to anticipate consumers' willingness to pay for such systems. This chapter
argues that a goals-achievement matrix may be one of the best evaluation
techniques for Intelligent Transportation Systems, such as an ITS-enhanced
transit system.

APPENDIX: APPLICATION OF THE CHECKLIST

Category	Goal[a]	Indicator	Effect	Comment
Congestion and delay	3(d), 3(e)	Peak-hour VMT	—	—
	3(d), 3(e)	Vehicle-hours of delay	—	—
	3(d), 3(e)	Volume/capacity ratio	—	—
	3(d), 3(e)	Peak/non-peak travel times	—	—
	3(d), 3(e)	Average speed (e.g. PMT/P-Hr)	—	—
	3(d), 3(e)	Average length of queue	—	—
Capacity and mobility	3(c)–(e)	Ease of personal mobility	—	For economically and physically disabled
	3(c)–(e)	System accessibility	—	(same as above)
		Network or connectivity implications	—	—
	3(c)–(e)	Transit passenger ridership	—	—
	3(d), 3(e)	Daily VMT	—	—
Safety and accidents	1(c), 2(b), 2(d)	Deaths	—	—
	1(c), 2(b), 2(d)	Injuries	—	For bus drivers and passengers
		Property damage	No Effect	—
Cost relative to other options		Implementation		
	2(a)	Operations	Uncertain	—
	2(a)	Maintenance	Uncertain	—

Category	Goal[a]	Indicator	Effect	Comment
Financial and fiscal outcomes	2(a)	Revenue enhancement	—	Due to: increasing service
	2(a)	Budgetary sufficiency	—	(same as above)
		Inter-governmental transfers	—	ADA requirements
		Private profits	—	—
Environmental outcomes	3(f)	Total emissions	—	—
	3(f)	Ground-level ozone	—	—
	3(f)	Energy consumption	—	—
		Hazardous materials impact	No effect	(Same as above)
		Soil erosion	Uncertain	Prevents the need for more highway construction
		Wetlands and habitat	—	(Same as above)
Economic outcomes	3(d), 3(e)	Gross regional product	Uncertain	—
		Employment	Uncertain	Increased employment: demand (e.g. dispatchers)
		Per capita income	Uncertain	Decreases (bus drivers – more)
		Land values	Uncertain	—
		Regional competitiveness	Minimal effect	—
Equity outcomes	3(a)	Affordability	Uncertain	If there is a change in fare

Category	Goal[a]	Indicator	Effect	Comment
		Regional access disparity	—	Better intermodal connection (bus to VRE)
	3(a)–(c)	Handicap access	—	—
Institutional outcomes	2(a)	Flexibility, adapts	—	Can add other ITS technologies
		Regional significance	Uncertain	—
		Promote synergy	—	Between PRTC and home services
		Intergovernmental cooperation	—	—
	3(b)	Community involvement and preferences	—	System is flexible; can accommodate different types of service (e.g., different routes)
		Ownership Changes	No effect	—
		Public–private ptnrs	—	—
		Liability	Uncertain	—
		Privacy	—	—
Spatial outcomes		Land use implications	—	May discourage high density development
		Land values	No effect	—
		Residential proximity issues	—	(Same as above)

Category	Goal[a]	Indicator	Effect	Comment
		Infrastructure design or geometry impact	—	May facilitate further development of "suburban" street layout
Demonstration		Transferability	Not very	Assumes that a flexible-route transit system is in place
		Value as a model	—	(Same as above)
		Suitability	—	—

Note: [a] The goals relate to Objectives 1–3 and their subobjectives described on pp. 135–6.

NOTES

1. Significant commercial development in Prince William County occurred since this evaluation was conducted in 1996. Dominian Semiconductor and AOL have both opened large
   facilities in the past two years.
2. Personal interview with Eric Marx of PRTC, September 1995.

REFERENCES

Belair, Robert, Westin, Alan and Mullenholz, A. (1993), *Privacy Implications Arising
from Intelligent Vehicle-Highway Systems*, Report to US Department of Transporta-
tion, Federal Highway Administration, December 8.
Chang, Shyue Koong and Schonfeld, Paul (1991), "Optimization models for compar-
ing conventional and subscription bus feeder services", *Transportation Science*
25(4), 15.
Dickey, J. (1983), *Metropolitan Transportation Planning*, New York: Hemisphere Pub-
lishing.
Fielding, G. (1987), *Managing Public Transit Strategically*, San Francisco, CA: Jossey-
Bass.
Kuhl, Mary (1995), *Improving Interbus Transfer with Automatic Vehicle Location*,
Year 2, Final Report (project funded by IDOT and the Midwest Transportation
Center).
Klein, H. (1995), "Evaluation case study: Montgomery County Advanced Transporta-
tion Management System (ATMS)", first draft.
Marx, Eric (1995), "PRTC's OmniLink serves surburban neighborhoods", *Passenger
Transport* October 6, p. 72.
Metropolitan Washington Council of Governments (1998), *Growth Trends to 2020:
Cooperative Forecasting in the Washington Region*, Publication no. 98817. http://
www.mwcog.org.
"SaFIRES: Virginia op-test to evaluate flexible routing service" (1995), *ITS APTS
Quarterly* **1**(3), 143.
Smith, O.L. (1991), "Computer-assisted flexible routing of mass transit systems", *Trans-
portation Quarterly* **45**(4), 581–97.

8. The Woodrow Wilson Bridge

Kingsley E. Haynes, William M. Bowen and Laurie Schintler

INTRODUCTION AND BACKGROUND

The purpose of this chapter is to create metrics for evaluating benefits and costs related to the use of Intelligent Transportation Systems (ITS) on the Woodrow Wilson bridge (WWB), and to provide transportation authorities with tools for assessing ITS funded projects.

The Woodrow Wilson Bridge, located in the US National Capital region, is a drawbridge and the only major Potomac River crossing on the south side of the National Capital Beltway (I-495). It is a major commuting route linking Virginia (VA) and Maryland (MD) as well as a major link in the north–south US East Coast I-95 Corridor. The current six-lane WWB is in poor condition and is being replaced with a higher capacity 12-lane bridge. In this chapter, the analysis focuses on a high bridge and advanced technology replacement vision. Not only is this the most probable solution to be adopted but the effects of technology on a high or low bridge are practically the same. As a benchmark, the chapter suggests evaluation procedures for the high technology bridge alternative.

Recent History of the Woodrow Wilson Bridge

The Woodrow Wilson Bridge faces serious problems that range from basic engineering issues, such as structural integrity, to operational issues such as its role as a "bottleneck" generating long periods of congestion, and as a source of increased accident/incident rates. Federal, state and local authorities are planning to build either a low or high bridge near the location of the existing bridge. Other alternatives (for example, a tunnel or other bridges to the north or south) were not adopted because of geographic and cost considerations, local and community concerns, and the unique traffic demand patterns associated with the present bridge.

The Woodrow Wilson Bridge is an essential part of the I-95 Corridor,

National Capital Beltway (I-495), and is an integral link in the Washington metropolitan transportation system. Due to heavy traffic, its structure is seriously jeopardized, requiring urgent measures to protect integrity in the short run and solve bridge-related infrastructure problems in the long run. The proposed long-term solutions include the construction of a new bridge near the existing one, or construction of a new bridge 2500 feet south of the existing site. The construction of a tunnel was discouraged by the nearby Alexandria community due to heavy pollution emissions expected from high chimneys needed to ventilate the tunnel interior (JHK Associates, 1995).

Growth in the region is expected to result in increased demand for bridge use. The Washington Council of Governments (WCOG) estimates that employment in the region is expected to grow by 50 percent, and population by 40 percent from 1990 to 2020. The bridge, which is now 34 years old, is slated for reconstruction, to be completed by 2010. Two parallel drawbridges – 25 feet higher than the current bridge – are planned. The added height eliminates the need to raise the drawbridge for most pleasure boats that pass under the bridge during peak periods. Most commercial marine traffic, which requires the drawbridge to be raised, pass under the bridge during non-peak hours.

Tolls are a near certainty on the new bridge. WCOG, which is responsible for regional transportation policy, has stated that tolls are necessary to offset the estimated $2 billion cost of the new bridge. The bridge is expected to have 12 lanes after construction, and the Beltway is scheduled to be widened to 10 lanes after construction to accommodate increasing traffic volumes. Nevertheless, due to the expected growth in the region, congestion may continue. Therefore, the possibility of charging varying fees for bridge use may reduce congestion by altering route choices or departure times, and increase revenue by making toll collection more efficient. The introduction of Intelligent Transportation Systems to provide information to drivers and automated variable tolling potentially increases the efficiency gains of such a system. In addition, accident rates on the Woodrow Wilson Bridge are twice those of the rest of the Capital Beltway. The introduction of ITS systems – capable of accident reduction through increased driver information and warning signs, as well as monitoring for aid in case of accidents in order to reduce response times – is also a possibility.

Traffic is heaviest during rush hours when commuters travel to and from work. Although the majority of vehicle traffic is from commuting, precise figures are not available. Vehicles from out of the bi-state and district region are mainly trucks. The Woodrow Wilson Bridge is a major link on the north–south US I-95 East Coast Corridor. Estimates of future total flow are 250 000 vehicles per day for both in and out-of-state (USDOT, 1996b).

The range of ITS technologies available for use is large: new technologies are evolving and benefits often overlap. Below, specific traffic management

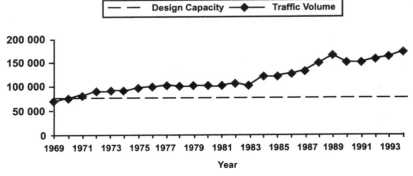

Source: USDOT (1996)

Figure 8.1   Traffic volume on the Woodrow Wilson Bridge, 1968–94

goals are reviewed as they relate to the Woodrow Wilson Bridge and specific ITS technologies. The multiple benefits of ITS technologies are seen as a positive. While these goals are represented separately, it is important to note that there is partial overlap in the benefits.

Current Conditions of the Woodrow Wilson Bridge

The Woodrow Wilson Bridge corridor is a five-mile stretch of highway that runs from Telegraph Road in Virginia (VA) to Indian Head Highway in Maryland (MD). The design capacity of the bridge is 75 000 vehicles per day; the bridge is a class C highway, designed to carry 1550 vehicles per hour per lane (1550 x number of lanes = 9300 per hour) (Transportation Research Board, 1985). However, as shown in Figure 8.1, traffic volume on the bridge has grown to twice the designed capacity. Figures 8.2 and 8.3 illustrate the origin and destination of commuting trips in 1990.

Eastbound traffic goes predominantly to Washington, DC and Prince George's County, MD, while westbound traffic goes to Fairfax, Alexandria and Arlington County, VA. Average daily traffic volumes are as follows: a.m. peak hour 12 140 and p.m. peak hour 12 590. A significant portion of this traffic is heavy truck traffic on the I-95 corridor, a major contributor to traffic congestion and weight-related stress to the bridge.

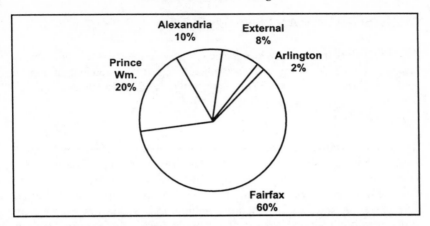

Figure 8.2   Origin and destination of commuting trips in Virginia in 1990

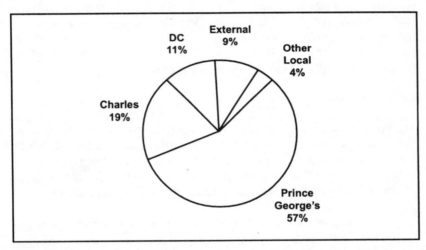

*Figure 8.3   Origin and destination of commuting trips in the Maryland
portion of the DC metropolitan region in 1990*

CONGESTION MANAGEMENT

A variety of complementary ITS traffic congestion management options are possible for the Woodrow Wilson Bridge. Congestion management can be applied to three elements: traffic diversion, tolling, and information to drivers.

Diversion of traffic

Re-routing of traffic is a sound way to manage congestion, especially when users save time and do not incur a sharp increase in time or distance costs. This is particularly attractive on the Woodrow Wilson Bridge for long-distance commuters and interstate travelers. Because alternative routing on the Beltway around Washington, north or south, is less than 5 miles, one way to divert traffic is through the use of electronic message signs posted well in advance, allowing (or reinforcing) alternative route utilization by supplying congestion information to drivers. A list of ITS technology options related to traffic diversion or re-routing includes electronic message signs, in-vehicle communication, radio communication, lane control, variable speed advisory (speed limit warning signs for each lane), and traffic lights (using red, yellow and green lights on specific lanes to manage flow).

Automatic tolling

Automatic tolling is another tool for congestion management because stop-and-go slow-downs are avoided and, if desired, congestion pricing can be implemented efficiently. Automatic tolling acceptability among users is important, but, of course, needs to be assessed against other benefits. The technologies and information required to implement effective electronic tolling are:

- vehicle-based transponders (with at least read and possibly read and write capabilities);
- infrastructure-based sensors to interact with vehicle-based transponders;
- information management and centralized systems for management of transactions to maximally reduce costs;
- congestion pricing analysis with respect to income, business expenses, alternative travel modes available and alternate routing possibilities.

Central to this are studies of price elasticities of demand which are used to assess the level of toll prices necessary to provide an incentive for a portion of drivers to switch driving times or modes (for example, shift to public transportation, carpooling, or rescheduling travel). Reaction to higher toll pricing is not clear; current studies to assess this require evaluation.

Information to Drivers on Congestion

Message signs and other instruments are useful in providing information relevant to avoiding the problems delineated in the previous sections. Warning drivers of congestion or accidents ahead is important to reduce stress, to avoid an increase in congestion, to reduce the potential for accidents, and to provide drivers with alternative routes. The applicable ITS technologies are: ·

- infrastructure-based sensors to manage end of queue warnings;
- video cameras for accident/incident identification and response assistance;
- variable message signs, radio broadcasts, or closed communications to provide information on accidents, congestion, alternative routing and lane change requirements.

Flow Improvement

Improving basic traffic flow reduces congestion, thereby decreasing travel time, delays, accidents and pollution. The following technologies improve flow control:

- HOV lanes on bridge and approaches as part of an integrated Beltway system;
- carpooling and vanpooling facilitation through increased information and priority to participating vehicles.

In addition, improving transit ridership through the use of variable pricing – which encourages peak hour use – may also reduce congestion. Here, there is an interesting potential interaction between congestion (peak) pricing for highways and variable (possibly reduced) pricing on complementary mass transit systems.

Safety and Incident Management

Incident management is an important element of congestion reduction. Traffic incidents and accidents are a primary cause of congestion, and information technology measures to minimize their number are of central concern. Further technologies that detect and reduce accident response times minimize the time needed to clear an accident scene. Diverting traffic from accident sites also reduces safety related sequencing problems (for example, accident related queuing, rubbernecking and life-saving by rapid trauma management). The main ITS technologies related to these concerns are:

- real-time management systems with communication systems such as message signs for re-routing traffic;
- variable speed advisory communications for each lane;
- volume control through automated lights to regulate traffic flows;
- continuous video monitoring of the bridge and its approaches for improved accident response.

Cost Recovery

Automated tolling will improve the efficiency of cost recovery on the WWB due to supply issues of higher flow leads, higher throughput capacity (due to lower congestion), and faster average speeds, and lower costs (relative to manual tolling) in operating the system (lower transaction costs). Because of the interjurisdictional context of the Woodrow Wilson Bridge, expenses and cost recovery will be shared among the partners. Issues to consider in association with automated tolling technology include:

- privacy issues regarding transponder transactions;
- investment obstacles, such as public resistance to smart-card technologies and tolls;
- transponder performance in high traffic volumes, at high speeds, or in high radio wave use areas.

Automatic toll collection depends on some kind of automatic vehicle identification equipment. There are three basic technological requirements. The first is the transponder; there are two types of transponders: read only, and read/write. Evaluations of this technology indicate that success is dependent on the free provision of transponders. Many transportation systems provide transponder equipment for free.

The second requirement is an infrastructure-based reading device located along the highway to identify passing vehicles. The third requirement is a transaction management mechanism that correlates toll accounts, vehicle passages and so-called "back-room" activities, as well as evaluating the impact of toll cost on flows given that alternate routes are available. The latter evaluation function may be important even if alternative routes are not available to assess price, total demand and time demand, elasticity and equity considerations.

Evaluation Dimensions

Temporal perspective

Ex post and *ex ante* evaluation of transportation investment decisions are important for policy, planning and management considerations. Prior to a project's deployment, as much information on the impact of transportation investment should be collected, not only to ensure wise allocation of public resources (for example, cost/benefit analysis), but also to assess public reaction and assessment. Further follow-up studies are needed to ensure fine-tuning of the prior investment and to provide lessons for further investments (Haynes and Li, 1993). It should be noted that with ITS technologies, the first capital IT investment is not the end; maintenance operations and support, as well as joint operations, represent a significant continuing investment.

Information generation

An important part of transponder use is the generation of information that can be used when designing improvements or in making evaluations. Transponders constitute an important source of information about highways, which can be used to increase the efficiency of their use. Statistical data on highway systems and ITS on highways are difficult and expensive to gather. The use of ITS may aid in collection of relevant information, and the building of real-time traffic management models that help guide response times to both congestion and accidents (for example, neural networks and learning management systems). Billing information for electronic tolls may also provide helpful information. These mechanisms (as described above) include: ramp metering to collect information on the number of cars, and bridge congestion issues; two-way radio frequency transponders (to identify return traffic); and flow detectors to count cars (for example, loop systems and audio/video systems).

METHODOLOGIES

In order to evaluate the effects of ITS technologies on the Woodrow Wilson Bridge, a checklist has been used to organize the analysis (see Chapter 1). The checklist consists of different categories with indicator variables and the related goals associated with each indicator. The objective is to assess the effect of intelligent transportation technologies on each indicator for each category. The checklist contains broad categories which might be more or less relevant depending on the ITS case. An important consideration is that the effects may or may not be comparatively commensurate. Consequently, although specific data elements are used, the checklist is essentially a qualitative analysis using informed judgments.

Checklist Summary and Overview

For this case study, three checklists were produced for three selected ITS technologies applicable to the WWB. These three technologies are: (a) video monitoring; (b) automated tolling; and (c) information devices for drivers related to congestion and incidents. Comments below are limited to these three categories although others are presented in Appendix 1. The effects of each ITS technology on the WWB related to each category may vary. However, one complete checklist summarizing the effect of these technologies on the bridge is provided in Appendix 1.

The use of ITS on the Woodrow Wilson Bridge, particularly electronic *vis-à-vis* manual tolling, will increase travel flow significantly. All indicators in the "Congestion and delay" category are improved as flow should increase with the use of electronic/automated tolls. Capacity and mobility are intrinsically linked to congestion and delay, and as congestion decreases, capacity will increase significantly.

The two other ITS technologies, driver information devices and video monitoring related to congestion and incidents, are particularly important to improved safety and accident avoidance. Also, as with electronic toll technology, another benefit of these technologies is decreased congestion and delay.

The goals in the checklist should be considered as general and broad goals relating the purpose of the deployment of ITS technologies to the WWB. This is in contrast to the goals, related directly to the nine potential benefits used in our subsequent trade-off analysis, which include: (a) improved safety; (b) better traffic flow; (c) lower travel cost; (d) better environmental quality; (e) increased business activity; (f) faster construction cost recovery; (g) greater user acceptance; (h) better travel information; and (i) better planning information.

ALTERNATIVE EVALUATION METHODOLOGIES

In this section of the chapter, three evaluation methodologies are presented and applied to the Woodrow Wilson case to better examine the effects of using ITS on the Woodrow Wilson Bridge. The first methodology, probabilistic multidimensional scaling algorithm (PROSCAL), is used for assessing goals and preferences (Mackay and Zinnes, 1988). The second methodology evaluates what is known about pricing and traffic demand (elasticity) by examining a variety of studies on this topic and incorporating them into a meta-analysis. The third methodology uses an evaluation mechanism to assess the impact of pricing and congestion on the timing and route choice behavior of potential Woodrow Wilson Bridge users through an application of the bottleneck model.

Probabilistic Multidimensional Scaling Algorithm (PROSCAL)

PROSCAL requires the collection of information using an interview schedule and a set of interview questions that respondents use to reveal their choices (in a pairwise manner) between alternative benefits. Officials from the Maryland, Virginia, and the District of Columbia Departments of Transportation, the Council of Governments, and State Highway Police were surveyed to discover what implications they felt ITS technologies have for the Woodrow Wilson Bridge. Interest focused on surfacing any unforeseen or underlying problems.

Meta-analysis

Meta-analysis focuses on the implications of potential ITS pricing of the use of the Woodrow Wilson Bridge. Meta-analysis is defined by Button (1994) as "the use of formal statistical techniques to sum up a body of separate but similar studies". Meta-analysis allows a series of different and independent studies to be drawn together in order to help understand the impacts of expected policies. Meta-analysis goes beyond other narrow forms of analysis by creating a cross comparison of the results of a series of supposedly similar and parallel studies. This is particularly useful for evaluating the expected impacts of policies, at least to the extent that individual studies can provide generalized information. Here, the meta-analytical methodology is used to assess the various studies on price elasticities of demand and their potential impact on bridge use.

Bottleneck Methodology

One major concern – how to pay back the costs of building a new bridge – raises the question or possibility of tolling. There are two major types of toll pricing. One is a fixed toll for all times of the day, and the other is a variable price (peak price) toll; both methods are impacted by the price elasticities of demand. Bottleneck modeling (Arnott *et al.*, 1991) is used to evaluate the relationship between toll levels and their impact on bridge use. Specifically, the impact of the use of an alternate route (for example, the Beltway through Maryland) and the role of information on tolls and use levels are examined.

For this application of the bottleneck model, traffic congestion generates a queue behind a bottleneck. In this model, an individual's departure time is an endogenous variable (determined with the model). The model specifies commuters traveling from home to work passing through a bottleneck with fixed flow capacity.

Due to physical limitations at the bottleneck, some commuters will arrive at work on time and others will be late or will have to adjust their home departure time in order to arrive at work on time. The costs associated with late or early arrival are called schedule delay costs. These costs, in the Woodrow Wilson Bridge case, include travel time, schedule delay, and toll price (which can be fixed or variable). Within this system, each individual decides when to depart from home so as to minimize trip price.

Probabilistic Multidimensional Scaling Algorithms (PROSCAL)

In evaluating the use of ITS on the Woodrow Wilson Bridge, it is important to keep in mind that different technologies may be applied to the bridge, and that decision making should be linked to the potential benefits ITS may bring to its users. This raises the question of goal or benefit priorities, something that can be addressed by using traditional decision utility models and (more recently) multi-dimensional scaling algorithms.

Although the ordering of collective preferences is not easy, it can be achieved or supported through the application of methods such as probabilistic multidimensional scaling algorithms (PROSCAL). This method aggregates individual preferences measured through survey techniques into collective or aggregate preferences which can then be used to help select appropriate ITS methods to address the top preference priorities (Mackay and Zinnes, 1988).

The PROSCAL methodology
The use of ITS in transportation infrastructure is characterized by risk and uncertainty. One of the consequences of this is a lack of agreement on ITS issues. Moreover, because of ITS' relative novelty, there are few empirical studies and real-world applications to refer to. Therefore, there is no a priori agreement on what potential ITS benefits to the Woodrow Wilson Bridge have the highest priority.

PROSCAL is used in this study to evaluate the results of a survey of transport officials on their views about the utilization of ITS technologies on the Woodrow Wilson Bridge. It consists of evaluating priorities related to nine competing goals of the project where collective consensus does not exist. Prioritization is determined from respondents' answers to a survey consisting of a list of pairwise goals. The nine goals are combined in a total of 365 pairwise comparisons. Each comparison has a grade ranging from a maximum preference scale of 4 on each side of the pair to a neutral preference of 0 (see Figure 8.4). The (PROSCAL) methodology calculates the distance of each goal to an ideal point and through the analysis of the standard deviation of these distances collective preference ordering is achieved (Bowen and Haynes, 1994).

PROSCAL's approach to multi-attribute analysis is achieved through a combination of Thurstone's (1927) theory of comparative judgment, and Heffner's

(1958) probabilistic assumptions. Monte Carlo simulations conducted in preparation for this study showed that under the assumptions made by the conventional utility approach, PROSCAL yields identical collective preference orderings. However, in reality, this occurs only when: (a) the value structure of the respondents is shared and fixed; (b) no information is lost or gained by either mode; and (c) the weighted additive model is used. Furthermore, PROSCAL has been thoroughly tested and compared favorably with several better-known decision models (MacKay and Zinnes, 1988). Empirical differences between collective preference orderings estimated by the conventional utility approach and the cognitive-PROSCAL approach arise only when the economic/behavioral assumptions of the conventional approach are relaxed, as in most real world decision situations.

Maximum likelihood estimates are used to approximate the model's location and variance parameters. Hypothesis tests are used to identify the model that best fits the data. The problems of evaluating the likelihood function along with further details of the model are considered in Zinnes and MacKay (1987).

Having estimated and identified the best-fitting model for a set of judgments, the collective preference ordering may be determined by either of two techniques. The first determines collective preferences by ranking the Euclidean distance between each alternative and the single ideal point. Greater dis-

Most preferred	**Neutral**	**Most preferred**
Improved safety 4 3 2 1	0 1 2	3 4 Better traffic flow

Improved Safety: Technologies such as lane video monitoring and congestion and accident warning signs, among others, increase safety and reduce accidents. These include improving conditions which decrease the probability of accidents on the bridge and/or increase response time when accidents occur.

Better traffic flow: Technologies such as lane change advisories and electronic versus manual toll collection, among others, improve traffic flow and reduce commuter time. These allow an increased number of vehicles (per hour) to use the bridge

Note: Respondents were instructed to circle the option that most accurately reflects their preference between the two alternatives, with 0 indicating a neutral preference and 4 indicating their maximum preference. The questionnaire contained 36 of these paired comparisons between all possible pairs of 9 judgments for each participant.

Figure 8.4   *Example of priority benefit preference for use of ITS on the Woodrow Wilson Bridge*

tances imply stronger preferences (higher priorities). The magnitude of the variance on each alternative is interpreted to reflect the relative degree of consensus on the preference for that alternative. The disadvantage is that this technique may require a secondary decision process, independent of the priorities, to consider explicitly the consensus issue.

The second technique determines the priorities by ranking the expected distances between each alternative and the ideal point. But in this case, because the collective preference estimates implicitly account for the consensus, they may not be monotonic with those determined under the first technique. Larger variances on an alternative tend to increase the expected distance between that alternative and the ideal point, which tends to increase its priority. The advantage of this technique is that it systematically factors consensus into the collective preference estimates, making interpretation easier and avoiding the need for any secondary decisions. The disadvantage is that it does not allow explicit estimation of consensus independent of priority.

Below, the survey information, the questionnaire, the analysis of results, the qualitative information self-assessment responses, and conclusions are presented.

Transportation questionnaire

The survey contains a description of the nine ITS benefit goals; a survey of the relative preferences for these benefits (36 paired comparisons); and a set of self-assessment questions.

The survey was performed to evaluate preferences of ITS technology benefits by transportation officials for the Maryland, Virginia, and District of Columbia Departments of Transportation, USDOT, the Council of Governments, and State Police. These officials were asked to make pairwise comparisons to determine the relative value they feel ITS technologies have for the Woodrow Wilson Bridge.

The survey consists of a questionnaire containing nine potential ITS benefits to the Woodrow Wilson Bridge. The first seven benefit goals adopted from Perera (1992), and USDOT (1995) are improved safety/accident reduction; better traffic flow; lower travel cost; better environment quality; increased business activity; faster construction cost recovery; and greater user acceptance. These are the goals of the National ITS Program plan (USDOT 1995). The other two adopted from USDOT (1996a) are: better travel information, and better planning information. These appear to be the potential benefits of ITS for transportation infrastructure. The nine potential benefit goals were defined in the survey as follows:

1. *Improved safety/accident reduction* Technologies, such as traffic video monitoring and congestion and accident warning signs, among others,

increase safety and reduce accidents. These include improving conditions which decrease the probability of accidents on the bridge and/or increase response time when accidents occur.

2. *Better traffic flow* Technologies such as lane change advisories and electronic versus manual toll collection, among others, improve traffic flow and reduce commuter time. These allow an increased number of vehicles (per hour) to use the bridge.

3. *Lower travel cost* Technologies such as early warning signs for congestion, reduce commuting and/or operating costs for both commercial and non-commercial vehicles.

4. *Better environmental quality* Technologies, such as automatic toll collection and early warning signs for congestion and alternate routes, have positive environmental impacts for local residents. These include, but are not limited to, reduction of noise, congestion, pollution and other traffic-related quality of life factors.

5. *Increased business activity* Technologies such as early warning signs for congestion, alternate route guidance information and electronic toll collection positively affect the transportation/routing times of commercial vehicles and open area businesses to new customers. These include technologies which may provide new employment or increase productivity.

6. *Faster construction cost recovery* Technologies such as electronic toll collection allow low cost recovery of construction costs used for bridge construction and renovation.

7. *Greater user acceptance* Technologies such as video monitoring, smart cards, and route guidance are to some degree acceptable and appreciated by the public, but others consider them an unacceptable intrusion on privacy and an annoyance.

8. *Better travel information* Technologies such as automatic vehicle location and travel information systems, especially for transit, provide real-time information including arrival times, departure times and delays.

9. *Better planning information* Data received through technologies such as automatic counters may be used by traffic management as additions on corrective action, future planning and scheduling, or reporting purposes.

From these nine benefit goals, 36 pairwise comparisons were generated and set in an optimum order [$n(n-1)/2 = 36$]. The order of the paired comparisons is important to eliminate errors, avoid influences on judgment, and decrease fatigue effects (Ross, 1934).

Survey results

The questionnaire was sent to a non-random list of 85 transportation professionals along with five extra questionnaires (on average) to be passed along to other colleagues involved in transportation planning. In total, 36 were returned. Of the non-responding officials, some stated that they did not feel directly involved with issues related to the Woodrow Wilson Bridge, and others felt that the questions were tiresome. Yet there was a 100 percent return from the Federal Highway Administration (FHWA), probably due to the covering letter stating that the survey was being carried out in close cooperation with FHWA.

Of the 36 questionnaires received, four were not utilized: one was received after the analysis was completed; two responded all 4s; and one was dropped from the analysis because of weak transitivity.

The results are summarized in Table 8.1: column 1 is a list of the nine benefits; column 2 is the combined collective preference ordering, computed by the "expected distance technique"; column 3 presents the non-combined collective preference ordering computed by the "Euclidean distance technique"; and the last column (4) presents the level of agreement.

Table 8.1   Collective preference ordering and measure of consensus for nine ITS benefits

Benefits	Priority		Agreement level
	Expected distance technique	Euclidean distance technique	
Improved safety	1	4	0.315
Better traffic flow	2	3	0.644
Lower travel cost	5	5	8.993
Better environment quality	3	2	4.183
Increased business activity	8	8	24.823
Faster construction cost recovery	9	9	34.301
Greater user acceptance	7	7	14.300
Better travel information	4	1	6.318
Better planning information	6	6	13.664

The results of the group judgments (summarized in Table 8.1) follow for each of the nine benefits in the Woodrow Wilson Transportation Survey:

1. *Improved safety/accident reduction* is a high priority goal regardless of how it is considered. Using the Euclidean distance technique (column 3) to estimate the priority, this goal is the fourth priority. If, however, the

second approach to analysis is adopted where part of the utility of a goal is considered to be inherent in the expert's agreement of its importance, then improved safety and accident reduction is the first priority (column 2). This shift of priority is due to the high level of agreement on its importance (column 4).

2. *Better traffic flow* is similar in some respects to improved safety and accident reduction. It is also a high priority goal, in one way or another. As with improved safety and accident reduction, it also has a high enough level of agreement to improve its priority when interpersonal agreement is considered.

3. *Lower travel cost* is a mid-level priority goal, regardless of how one looks at it, though in comparison to the first two goals there is quite a bit of disagreement about the importance of lower travel cost. However, the disagreement does not affect its priority.

4. *Better environmental quality* is the second highest priority goal if priorities are estimated using individual aggregates, without considering cross-subject agreement level. Moreover, agreement is high in comparison to all but the first two goals. However, agreement is not high enough to raise this goal's priority as in the case of the first two. In the case of better environment quality, the level of agreement slightly reduces the priority from second to third.

5. *Increased business activity* is, overall, a low priority and a controversial goal. This result does not vary with the technique used for the priority estimation.

6. *Faster construction cost recovery* is the lowest priority goal, and is also the least agreed upon. Evidently, some of the experts consider it to be quite important; others not at all.

7. *Greater user acceptance* is a middle to low priority, with middling levels of agreement. The level of agreement does not affect its priority.

8. *Better travel information* is the highest priority goal for the group if one simply aggregates across the expert judgments disregarding cross-subject agreement. However, there is considerable disagreement between the experts on the appropriate priority for this goal. In fact, where an estimate of the value of the disagreement is considered, the priority for better travel information drops from first priority to fourth.

9. *Better planning information* is a low–middle priority with fairly high disagreement. The level of disagreement does not influence the priority estimate one way or the other.

Individual level analysis Distances mean disutilities. Larger distances imply greater disutilities, the smaller distances imply smaller disutilities (or great utilities). Table 8.2 presents the ranking of individual priorities using the expected and Euclidean distance technique. Although the distances vary depending on the technique used, the ordering is the same for both techniques. The rankings are based on the distances presented in Appendix 1 (the higher priorities are associated with smaller values or greater utilities). For example, Table 8.2 shows that most of the respondents preferred benefit number 8 (better travel information), regardless of the distance technique applied.

Self-assessment   In Appendix 4, tables are presented with basic information on the questions in the self-assessment part of the survey. From Tables 8A.1 and 8A.2 the majority of the respondents possess 16 years or more of transportation management experience, but no direct experience in ITS activities. Most of the respondents are transportation planning and analysis professionals who work in the District of Columbia. Most, however, are neither Woodrow Wilson Bridge commuters nor planners or researchers. There are likely to be significant differences among respondents based on experience in the transportation sector and source of professional employment.

Summary and conclusion of the PROSCAL analysis
Although conventional decision analysis based on the utility theory has its own merits, it has restrictive assumptions on the variation in judgment across individuals (Bowen and Haynes, 1994). On the other hand, multi-dimensional scaling is a useful and robust technique for assessing collective preferences.

The results of the PROSCAL algorithm differ depending on the technique used. With the expected distance technique, the priorities are in order: (1) improved safety; (2) better traffic flow; (3) better environmental quality; (4) better travel information; (5) lower travel cost; (6) better planning information; (7) greater user acceptance; (8) increased business activity; and (9) faster construction cost recovery. The ordering regarding the Euclidean distance technique is somewhat different: improved safety moves from first priority to fourth; better traffic flow from second to third; better environmental quality from third to second; and better travel information from fourth to first priority. These differences are due to the weight put on the role of consensus among experts.

This methodology, like the others in this paper (meta-analysis and the bottleneck model) is an important tool for evaluating benefits of ITS to transportation infrastructure. It identifies the ITS benefits that should be given priority based on preference level. The top four preferences for the ITS benefit goals are: improved safety; better traffic flow; better environmental quality; and better travel information, regardless of the PROSCAL method of analysis used. Less-preferred ITS benefits are: lower travel cost; better planning information; greater user acceptance; increased business activity; and faster construction cost recovery.

Table 8.2 *Individual level analysis: individual priorities by benefits*
         (expected and Euclidean distance technique)

Respondents	Benefits								
	1	2	3	4	5	6	7	8	9
1	4	3	5	2	8	9	7	1	6
2	4	3	5	2	8	9	7	1	6
3	5	4	6	3	8	9	7	1	2
4	5	4	6	2	8	9	7	1	3
5	4	5	3	6	2	1	9	7	8
6	5	4	6	3	8	9	7	1	2
7	3	2	5	1	7	9	8	4	6
8	3	2	5	1	7	9	8	4	6
9	4	3	6	2	8	9	7	1	5
10	6	5	7	4	8	9	1	3	2
11	4	5	3	6	1	2	9	7	8
12	4	3	5	2	8	9	7	1	6
13	4	5	3	6	2	1	9	7	8
14	7	3	9	2	1	5	8	4	6
15	4	3	5	2	8	9	7	1	6
16	6	5	7	4	8	9	3	1	2
17	3	2	4	1	6	7	9	5	8
18	4	3	5	2	8	9	7	1	6
19	4	3	5	1	8	9	7	2	6
20	4	5	3	6	2	1	9	7	8
21	5	4	6	2	8	9	7	1	3
22	3	1	4	2	6	7	9	5	8
23	4	3	5	2	8	9	7	1	6
24	3	2	4	1	7	8	9	5	6
25	4	3	5	2	8	9	7	1	6
26	3	2	4	1	7	9	8	5	6
27	5	4	7	3	8	9	6	1	2
28	4	3	5	2	8	9	7	1	6
29	3	2	4	1	7	8	9	5	6
30	6	5	7	4	8	9	1	3	2
31	6	5	7	3	8	9	4	1	2
32	4	5	3	6	2	1	9	7	8

Meta-analysis: Congestion Pricing and Price Elasticities of Travel Demand

The unique location of the Woodrow Wilson Bridge has resulted in serious traffic congestion and safety problems, particularly during rush hours. The 35-year-old bridge not only provides a convenient way for commuters to cross the Potomac River between the states of Virginia and Maryland, it also allows interstate travelers to bypass Washington, DC to the south. Moreover, substantial local and regional growth has increased the travel demand across the bridge to more than twice the original design capacity. The number of hours of congestion per day on the bridge has increased from one hour to over three hours during peak periods (USDOT, 1996b). Several proposed solutions for replacing this deteriorating structure including the application of Intelligent Transportation Systems (ITS) have been discussed in the Washington metropolitan region. The goal is to resolve the imbalance between capacity and demand of the bridge. The use of ITS technologies, particularly an automatic tolling system, would impact the current traffic flow and future levels of congestion.

ITS technologies, including automatic tolling, have been widely utilized on highways, bridges, and tunnels to facilitate traffic and congestion management for decades in the US and elsewhere. From one perspective, automatic tolling would allow the bridge not only to generate revenues to finance construction costs, it would also reduce operational costs. Automatic tolling can also function as a mechanism to encourage more efficient use of the bridge via appropriate congestion pricing schemes. With ITS technologies providing information to drivers, and automated variable tolls charging varying fees during peak periods, the number of vehicles crossing the bridge may decrease, resulting in reduced congestion and fewer safety problems. Theoretically, congestion pricing would cause bridge users to change their travel behavior based on the best times to use the bridge and the availability of other travel modes or routes. These choices would be made in order to gain an optimal travel cost.

The appropriate level of tolling and congestion prices for the bridge would be a central subject among decision makers, transportation professionals, and the general public. With an appropriate tolling or congestion pricing scheme, the volume of traffic could be reduced while still attracting a sufficient number of users to cover construction costs. Thus, understanding how tolls and congestion pricing could be used as an incentive to impact drivers' behavior would rely on evaluating associated issues (for example, price elasticities of travel demand, traffic flow effects, costs and benefits, and others). In this section of the chapter, interest focuses on measuring the impact of price elasticities on travel demand.

Price elasticity is one aspect of travel demand analysis conducted by economists and transportation planners in measuring how users respond to price change. Although there is a general consensus that an increase in tolls or the implementation of congestion pricing would reduce highway traffic volumes, the size of the difference is still an issue. Very few countries in the world have implemented congestion pricing even on an experimental basis. Hence, research done to evaluate its impacts on drivers' behavior is very limited. Economists, however, have examined numerous empirical data sets to determine the level of price elasticities based on an increase in tolls, fuel prices, or transit fares. Here, the meta-regression analytic technique suggested by Button (1994) is used to analyze empirical studies that evaluated the price elasticity of demand associated with such changes. Using this approach, an aggregate understanding of the relationship between the elasticity of price change on demand may be achieved.

Issues

Congestion pricing   The idea of congestion pricing is derived from the notion that automobile users should pay for the marginal cost of each trip they make in terms of its impacts on other users' entry and movement through the traffic stream. For public facilities (such as highways, bridges, and tunnels) with limited capacity, toll volumes would be higher during peak commute hours when demand is highest, and lower in off-peak hours when traffic does not exceed a facility's capacity. Varying pricing schemes are already common in private industry, such as for electric and telephone service and in the airline and hotel industries. Two characteristics are essential to congestion pricing: (a) tolls must vary widely by time of day in order to discourage peak-hour travel; and (b) the peak price must be high enough to influence travelers' behavior for congestion relief (Higgins, 1994; McMullen, 1993; Orski, 1992).

The issue of implementing congestion pricing in the public highway sector, however, has been controversial, particularly in the US. Advocates believe that a congestion fee will result in reduced traffic, enhanced highway user mobility, improved air quality, and increased energy conservation. Conversely, the perception that such fees are unfair, discriminatory, and anti-business has impeded the acceptance of congestion pricing by the general public. However, in recent years, worsening traffic congestion levels and improved ITS technologies have stimulated policy makers to rethink congestion pricing as a management tool.

Four alternative congestion pricing technologies have been used as direct or indirect methods of charging for road usage. They consist of:

- cordon pricing using manual tollbooths;
- supplementary licensing;
- off-vehicle recording systems, such as automatic vehicle identification

(AVI); and
- on-vehicle charging systems, such as smart cards.

With the application of ITS technologies, congestion pricing can be used not only as an incentive for more efficient use of road space but also as a potential revenue generator to finance transportation improvements. Electronic toll collection is now being used on many highways in the US and elsewhere. By applying this optional toll collection technology, the implementation of congestion pricing is just a technical step to be adapted to incorporate differentials between peak and off-peak traffic.

The practice of congestion pricing has been undertaken on a limited basis in a few European and Asian countries such as Norway, England, Sweden, the Netherlands, Singapore, and Hong Kong to control urban traffic. In the US the Intermodal Surface Transportation Efficiency Act of 1991 includes $150 million for five pilot congestion pricing projects in urban areas. The Transportation Efficiency Act of 2000 further supports the development of this approach.

Price elasticities of travel demand   It is recognized that congestion pricing does offer an alternative for policy makers to allocate existing road space, to manage traffic flow, and to reduce congestion without actually expanding highway capacity. Travelers would change their travel habits because of an increase in their peak-hour commuting costs. Similarly, changes in fuel prices and transit fares impact auto drivers' travel demand. For example, studies have found that, as a rough rule of thumb, transit demand declines by 0.33 percent for every 1 percent increase in transit fares. That is, the rate of decline corresponding to price elasticity is approximately 0.33 on average.

The price elasticities of travel demand serve as useful indicators to evaluate how travelers respond to changes in public policies that affect their travel costs. The principle is that the greater the elasticity of travel demand, the greater the reduction in travel volume resulting from an increase in toll price, and vice versa. Moreover, reduction in traffic volume will lead to less toll revenue. Several empirical studies conducted by Oum *et al.* (1992) and Goodwin (1992) measuring price elasticities for various transportation modes found that travel demand is inelastic in response to price changes. Their studies may indicate that price change is not the exclusive variable affecting travel demand. Other variables such as mode, time of day, travel purpose, household income, and the amount and direction of the price change may also have a significant impact on the measurement of elasticities. Similarly, evaluations in different years and at various locations may affect elasticity results greatly.

Better use of existing studies regarding price elasticities of travel demand with respect to different conditions can provide greater insight into this issue.

Moreover, adopting an aggregate approach to synthesize similar but separate empirical studies may be viewed as a forecasting model for the outcome of future empirical research. Meta-analysis is a technique that can be used to create such a synthesis. It employs statistical analysis to combine findings from similar empirical studies.

The Meta-analysis methodology

The employment of meta-analysis has a relatively long history. Since the turn of the century, the research methods in meta-analysis have been applied in the fields of medicine, psychology, and education research. In recent years, its use in social science research has shown rapid growth. The growing popularity of meta-analysis in the social sciences can be attributed to its ability to integrate and empirically explain the various results found in a set of individual studies as well as its ability to synthesize an increasing amount of information. It is a technique used to sum up the magnitude of some important parameters, to provide more accurate evaluation of these parameters, and to offer insights into phenomena that no specific study has currently examined (Button, 1994). In sum, the advantages of meta-analysis can best be found in Nijkamp's (1996, p. 2) description: "[meta-analysis is] a generic term for a spectrum of comparative analysis methods designed to generate additional information from an existing body of knowledge often incorporated in case studies or quasi-controlled experiments ... [It] comprises the systematic application of a range of statistical methods to assess common characteristics and variations across a set of actual separate, but similar case studies on more or less the same type of phenomenon."

Nijkamp (1996) further points out that the research methods involved in meta-analysis are often diverse. The reason is that when researchers conduct an analysis of various quantitative studies, the range of topics that can be addressed and the scope of the statistical issues that are covered are relatively wide. In addition to the conventional statistical method of regression analysis which is commonly used in meta-analysis work, he also synthesizes three other research methods: meta-multicriteria analysis, epistemological or expert analysis, and rough set analysis, providing an effective way of dealing with a much more diverse and less immediately tangible set of factors (see ibid., pp. 3–4).

Certainly, meta-analysis has its limitations and problems – particularly in the field of social sciences. Thus, the interpretation of research results should be undertaken with great caution. Button (1994) explicitly points out that, unlike in the medical or psychological fields where research results are reported in a fairly standardized form, styles of presenting research results in economics tend to be quite different. He mentions that the output of most traditionally conducted reviews tends to be in the form of taxonomies of findings without any specific attempt to relate these to the review's purpose. Moreover, the way researchers report their assumptions, the way data are collected with respect to meeting their study objective, and the methodology used to measure outputs are not standardized in economics. Although meta-analysis is used to reduce the subjectivity that exists in assessing different studies and to identify relevant common issues, the analyst should recognize that the problem of subjectivity can still be inherent due to the heterogeneous nature of data reported by individual studies. In sum, meta-analysis is known as a powerful tool in comparative studies, but its limitations also need to be recognized when the research method is undertaken. To define and seek to measure the extent of commonalities in background variables can be potentially problematic and inevitably subjective.

Data analysis and the model

The conventional statistical method frequently used in meta-analysis is meta-regression analysis (MRA). This model provides a tool for economists to summarize regression results across studies, given the fact that most individual economics studies tend to rely upon regression analysis for their hypothesis testing. A general function form of MRA for integrating empirical studies is:

$$Y_j = a + \Sigma b_k Z_{jk} + \mu_j \quad (j = 1,2,...N) \quad (k = 1,2,...M) \tag{8.1}$$

where:
Y_j represents a reported interest in the *j*th study from a literature of N studies;
a represents the summary value of Y;
Z_{jk} are variables of each empirical study measuring the M relevant characteristics that might explain variation across studies;
b_k are the regression coefficients of M characteristics;
μ_j is the error term.

In the case of the Woodrow Wilson Bridge, the MRA model is used to evaluate price elasticities of travel demand E as the dependent variable Y_j

Table 8.3   Meta-regression explanatory variables

Mode	Measurement	Facility
Auto	Tolls	Bridge
Total traffic	Fuel prices	Tunnel
Transit	Transit fares	Local roads
Light truck		
Heavy truck		

Sources: Gifford and Talkington (1994)

from six different empirical studies. The selection of empirical studies are somewhat subjective due to our focus on examining the relationship between congestion pricing or changes in tolls, fuel prices, transit fares, and drivers' behavior in using specific bridges, tunnels, or highway toll facilities. Within these six studies, three main categories of explanatory variables Z_{jk} – mode O, price measurement M, and facility F – are evaluated according to their correlation with the price elasticities of travel demand. Table 8.3 describes a list of all explanatory variables used in the MRA model. The general model tested is:

$$E = f(O, M, F) \tag{8.2}$$

where:
O = a vector of mode dummy variables with the total traffic as the base;
M = a vector of measurement dummy variables with the transit fares as the base;
F = a vector of facility dummy variables with local roads as the base.

The studies selected for meta-analysis are based upon secondary sources from both published and unpublished studies. The data collection process presents several constraints, and certain adjustments have been made in order to aggregate useful indicators into the research. First, not many empirical studies have been found that concern measurement of automobile drivers' behavior in response to the increase in toll prices. The topic of price elasticity of travel demand is commonly discussed, but not in terms of actually measuring the change of travelers' behavior. Meanwhile, with respect to the evaluation of congestion pricing impact, access to some unpublished papers (presented in European conferences) was limited: this constrained the sampling size of the research. Thus, the research was extended to include evaluations of price changes in fuel expenses and transit fares in order to gain a broader set of studies on how price fluctuation may affect users' behavior in different transportation modes. Table 8.4 shows the six studies to be analyzed.

Second, few studies have specifically been designed to measure and report price elasticities of travel demand. Studies that did not indicate the amount of price change or the volume of traffic change were excluded due to an inability to standardize the measurements. Some studies reported the change of traffic volume resulting from toll changes, as opposed to reporting price elasticities. Using the basic formula of price elasticities commonly used by economists, which measures the percentage change in quantity of a specific variable (for example, traffic volume) with respect to the percentage change in price, price elasticities were computed from those studies with data sets available on the change in traffic volume and tolls. This is a simplistic way to compute the

Table 8.4 Estimated price elasticities of travel demand

Study	Price elasticities[a]	Elasticity (standardized)[b]	Measurement	Mode	Facility
US DOT (1996b)	−0.086 ($1: 10 lanes)	−0.086	Toll	Auto	Bridge
McCarthy and Tay (1993)	−0.147 ($1.85)	−0.079	Congestion pricing	Total traffic	Local roads
	−0.047 ($1.85)	−0.025	Congestion pricing	Auto	Local roads
Wilson (1988)	+0.07 ($1.85)	0.038	Congestion pricing	Bus	Local roads
	−0.047 ($1.85)	−0.025	Congestion pricing	Auto	Local roads
	−0.023 ($1.85)	−0.012	Congestion pricing	Taxi	Local roads
Gifford and Talkington (1994)	−0.021 ($0.25 M–Thu)	−0.084	Toll	Auto	Bridge
	−0.187 ($0.75 Fri)	−0.249	Toll	Auto	Bridge
	−0.178 ($0.75 Sat)	−0.237	Toll	Auto	Bridge
	−0.048 (−$0.25 Sun)	−0.192	Toll	Auto	Bridge
	−0.122 (Mon–Thu)	−0.122	Fuel price	Auto	Bridge
	−0.222 (Fri)	−0.222	Fuel price	Auto	Bridge
	−0.093 (Sat)	−0.093	Fuel price	Auto	Bridge
	−0.084 (Sun)	−0.084	Fuel price	Auto	Bridge
Hirschman et al. (1995)	−0.26 ($1.75)	−0.148	Toll	Auto	Tunnel
	−0.07 (")	−0.04	Toll	Auto	Tunnel
	−0.13 (")	−0.074	Toll	Auto	Bridge
	−0.03 (")	−0.017	Toll	Auto	Bridge
	−0.09 (")	−0.051	Toll	Auto	Bridge
	0.19 (")	0.109	Toll	Auto	Bridge

176

Study	Price elasticities[a]	Elasticity (standardized)[b]	Measurement	Mode	Facility
	-0.50 (")	-0.286	Toll	Auto	Bridge
	-0.10 (")	-0.057	Toll	Auto	Bridge
	-0.54 ($1.75)	-0.308	Toll	Light trucks	Tunnel
	-0.45 (")	-0.257	Toll	Light trucks	Tunnel
	-0.07 (")	-0.040	Toll	Light trucks	Bridge
	-0.14 (")	-0.08	Toll	Light trucks	Bridge
	-0.13 (")	-0.074	Toll	Light trucks	Bridge
	-0.12 (")	-0.068	Toll	Light trucks	Bridge
	-0.17 (")	-0.097	Toll	Light trucks	Bridge
	-0.60 ($1.75)	-0.343	Toll	Heavy trucks	Tunnel
	-0.60 (")	-0.343	Toll	Heavy trucks	Tunnel
	0.00 (")	0.00	Toll	Heavy trucks	Bridge
	-0.14 (")	-0.08	Toll	Heavy trucks	Bridge
	-0.09 (")	-0.051	Toll	Heavy trucks	Bridge
	-0.01 (")	-0.006	Toll	Heavy trucks	Bridge
	+0.20 (")	0.114	Toll	Heavy trucks	Bridge
	-0.12	-0.1	Trans fare	Auto	Tunnel
	-0.02	-0.02	Trans fare	Auto	Tunnel
	-0.10	-0.10	Trans fare	Auto	Bridge
	0.23	0.23	Trans fare	Auto	Bridge
	0.07	0.07	Trans fare	Auto	Bridge
	-0.20	-0.20	Trans fare	Auto	Bridge
	-0.23	-0.23	Trans fare	Auto	Bridge

Study	Price elasticities[a]	Elasticity (standardized)[b]	Measurement	Mode	Facility
	0.08	0.08	Trans fare	Auto	Bridge
	0.14	0.14	Fuel price	Auto	Tunnel
	0.16	0.16	Fuel price	Auto	Tunnel
	0.07	0.07	Fuel price	Auto	Bridge
	0.07	0.07	Fuel price	Auto	Bridge
	0.05	0.05	Fuel price	Auto	Bridge
	-0.04	-0.04	Fuel price	Auto	Bridge
	-0.32	-0.32	Fuel price	Auto	Bridge
	-0.03	-0.03	Fuel price	Auto	Bridge
	-0.21	-0.21	Fuel price	Light trucks	Tunnel
	-0.18	-0.18	Fuel price	Light trucks	Tunnel
	0.08	0.08	Fuel price	Light trucks	Bridge
	-0.06	-0.06	Fuel price	Light trucks	Bridge
	-0.04	-0.04	Fuel price	Light trucks	Bridge
	-0.20	-0.20	Fuel price	Light trucks	Bridge
	-0.05	-0.05	Fuel price	Light trucks	Bridge
	-0.55	-0.55	Fuel price	Heavy trucks	Tunnel
	0.01	0.01	Fuel price	Heavy trucks	Tunnel
	0.09	0.09	Fuel price	Heavy trucks	Bridge
	-0.06	-0.06	Fuel price	Heavy trucks	Bridge
	-0.01	-0.01	Fuel price	Heavy trucks	Bridge
	-0.08	-0.08	Fuel price	Heavy trucks	Bridge

Study	Price elasticities[a]	Elasticity (standardized)[b]	Measurement	Mode	Facility
Frick et al. (1996)	−0.12	−0.12	Fuel price	Heavy trucks	Bridge
	−0.218 ($2)	−0.109	Congestion pricing	Auto	Bridge

Notes:

[a] Price elasticities calculated here have not been standardized.

[b] Standardized elasticities are calculated by dividing the original elasticity by the change in toll price. Original price elasticities for transit fare and fuel price were used because no data were available to standardize the elasticities.

data points of price elasticities without considering other exogenous factors, and it may generate bias later in the analysis.

Third, considering that the data points of price elasticities are measured by different levels of price change, standardization of the measurement of price elasticities is necessary. Thus, each data point of price elasticities is divided by the price fluctuation derived from each individual study. By doing so, it is possible to identify what level of traffic volume (or users) is impacted by increasing tolls, fuel expenses or transit fares by $1.

Finally, theoretically meta-analysis is an empirical exercise in which the data points are the individual studies, not individual observations. In our research, several studies, in fact, contain more than one evaluation of price elasticities. The study done by Hirschman *et al.* (1995) measures price elasticities of eight different bridges and tunnels in New York City. Thus we treat each measurement as a variable data point in this case instead of viewing the whole study as one data point. This model then uses 70 different estimations of price elasticities as data points.

Findings of the analysis

The results of the analysis are presented in two different forms: regressions of the original price elasticities; and standardized price elasticities. Table 8.5 lists the coefficients estimated from the original price elasticities regression. Although the significance of individual variables is relatively low in this model, a high T-statistic value indicates that explanatory variables included in this model significantly explain the estimated value of price elasticities at the 95 percent level of confidence interval. Looking at the coefficients, the negative sign and scale of price measurement variables (for example, tolls is –0.1086,

Table 8.5   Coefficients from original price elasticities model

Variable	Coefficient	Std error	T-statistic	Prob.
Intercept	0.0746	0.2354	0.3167	0.7526
Tunnel	–0.3302	0.2036	–1.6217	0.1101
Bridge	–0.1710	0.1975	0.8656	0.3901
Tolls	–0.1086	0.0675	–1.6082	0.1130
Auto	0.1000	0.1975	0.5062	0.6146
Light trucks	0.0377	0.2046	0.1843	0.8544
Fuel prices	–0.0087	0.0693	–0.1254	0.9007
Transit	0.1705	0.1975	0.8631	0.3915
Heavy trucks	0.0606	0.2046	0.2961	0.7682
Congestion pricing	–0.2216	0.1715	–1.2917	0.2014

$n = 70$, $R^2 = 0.2662$, T-statistic = 2.4189

congestion pricing is –0.2216, and fuel prices is –0.0087) provides confirmation of the relationship of price elasticity of travel demand and price changes as indicated in Gifford and Talkington (1994), who reported that the effect is inelastic within a small scale. These coefficients also indicate that the value of price elasticity of travel demand related to these three variables (tolls, congestion pricing, and fuel prices) is less than that for transit fares because their coefficient values are less than zero. This confirms the a priori expectation that an increase in tolls, congestion pricing and fuel prices would reduce the amount of traffic volume, while an increase in transit fares would not decrease the level of traffic congestion.

Moreover, lower coefficient values of heavy truck (= 0.023) and light truck (= 0.012) variables indicate that the value of price elasticity of travel demand related to trucks is less than other travel modes such as auto (coefficient = 0.054) and transit (coefficient = 0.092). This shows that an increase in travel costs would seem to affect travelers who use auto or transit more than users of trucks. Those who use public transit as their travel mode are particularly affected.

Comparing the relationship of the value of price elasticity and the facility variables (for example, bridges, tunnels, and local roads), one discovers that the value of price elasticity of demand related to bridges (its coefficient = –0.171) is less than for tunnels (= –0.3302). This suggests that the value of price elasticity affected by price changes would be reduced more when travelers use tunnels than bridges.

Similarly, Table 8.6 lists the coefficients estimated from the standardized price elasticities regression that shows relatively similar results for the relationship between the price elasticity of travel demand and respective explana-

Table 8.6   Coefficients from standardized price elasticities model

Variable	Coefficient	Std error	T-statistic	Prob.
Intercept	0.0183	0.2012	0.0911	0.9277
Congestion pricing	–0.0973	0.1466	–0.6641	0.5093
Bridge	–0.0840	0.1688	–0.1688	0.6206
Tunnel	–0.1823	0.1740	–1.0478	0.2991
Transit	0.0920	0.1688	0.5450	0.5878
Fuel prices	–0.0154	0.0594	–0.2586	0.7969
Heavy trucks	0.0232	0.1750	0.1328	0.8948
Light trucks	0.0122	0.1750	0.0695	0.9448
Tolls	–0.0477	0.0585	–0.8154	0.4182
Auto	0.0540	0.1688	0.3199	0.7502

$n = 70$, $R^2 = 0.1445$, T-statistic = 1.0884

tory variables. However, a low T-statistic value indicates that this model, in general, is statistically insignificant in explaining the estimated value of price elasticities at the 95 percent level of confidence interval. Looking at the coefficients of congestion pricing and tolls from the two models, both suggest that one unit of increase in congestion pricing has a higher negative impact on the price elasticity of travel demand than one unit of increase in tolls. This result confirms the perspectives of many congestion pricing proponents who advocate that the implementation of congestion pricing during the peak periods could reduce the level of traffic congestion more than an equivalent level of increase in tolls.

DISCUSSION

The focus of the Woodrow Wilson Bridge project (using meta-analysis) has been on forecasting the bridge users' responses to an increase in tolls. Although the methodology of meta-analysis applied in this case is an elementary exercise, it does provide a useful theoretical framework for future research: to make the best use of the information that is available. Evaluating the impact of policy changes on travelers' behavior is not uncommon, but building a complete array of key parameters – as well as various forecasting techniques on this issue – still needs to be done. From the meta-analysis process, it is recognized that it is useful for linking similar studies through statistical analysis to generate a broader perspective on a specific policy issue.

The results and summary statistics in the two models show that none of the independent variables emerged as statistically significant at any reasonable confidence level. One explanation is that the estimation is rather rough. If a wider range of explanatory variables could be derived from the studies selected, or if more studies were done regarding this issue, the MRA model might be able to show some interesting results. Another possible problem may be the bias resulting from the nature of the studies that are included. Certain research subjects are more easily published than others; it is hard to know whether significant differences exist in those studies that have not been accepted for publication. Therefore, how to establish explicit criteria for selecting the studies in any meta-analysis, and identifying the limitations of the selection process are challenges for any researcher.

Moreover, the calculation of price elasticity of demand can be a problem. The traffic volume used in the formula for calculating the price elasticity is only an estimate. Given that other relevant information (such as drivers' income and trip purpose) is not available in individual studies, the result of price elasticity may not accurately reflect the true impact of toll changes.

The lesson learned from this application of the meta-analysis technique is that a wider range of explanatory variables included in individual studies could reduce the probability of bias in the analysis. Particularly, factors such as the purpose of travel, commuting patterns, household incomes, and others should not be treated as constant variables in the formulation of measuring price elasticities of travel demand. Forecasting travelers' responses to travel cost increase is not solely determined by the difference of tolls. It is essential to consider those factors in the analytical process. Certainly, the deployment of ITS has an important role to play because it can help to obtain this information in a more efficient way. For instance, an advanced automatic tolling system could record detailed travel routes, commuting patterns, and travel destinations. In the future, these data will not only enhance the reliability of empirical studies, it will also provide policy makers with useful guidelines to improve congestion problems.

Bottleneck Analysis

This section provides two bottleneck model applications. The first demonstrates how a bottleneck method can be used to examine the effects of different policy scenarios relating to ITS and tolling on the Woodrow Wilson Bridge. Although this problem characterizes a real-life situation, the results derived from this simulation should not necessarily be used in drawing policy conclusions. This example is mainly intended to show how the bottleneck model could be used in policy analysis, particularly as it relates to Intelligent Transportation Systems (ITS) and tolling.

The second bottleneck application demonstrates how the characteristics of the bottleneck on the Woodrow Wilson Bridge can be identified in order to provide better congestion, incident, and accident management.

First bottleneck model application
The bottleneck methodology presented here follows Arnott *et al.* (1991), and presumes a travel time/schedule delay trade-off. There are three basic parts to the model, which characterize wait time, queue length, and departure/arrival rates to the bottleneck. These, in turn, affect the calculation of travel cost functions. Each component is described next.

Modal specification   Wait time at the bottleneck at any time *t* is described by the following relationship:

$$T_j(t) = D_j(t)/s_j \qquad (8.3)$$

where:
$T_j(t)$ is travel time on route *j* for a driver departing at time *t* ;

$D_j(t)$ is the number of vehicles in the queue on route j at time t ;
s_j is the flow capacity of route j (uncongested except for a bottleneck).

Queue length   Number of vehicles waiting at time t, is determined by the
rate at which vehicles enter the queue or the arrival rate, the number of vehicles already waiting in the bottleneck, and the capacity of the bottleneck. A
queue begins when the number of cars entering the bottleneck exceeds the
capacity. With a departure rate of $r_j(t)$ at time t, Arnott *et al.* (1991) give the
following function for the number of vehicles in the queue:

$$D_j(t) = \int_{t_{aj}}^{t} r_j(u)du - s_j(t - t_{aj}) \qquad (8.4)$$

where:
t is the actual departure time;
t_{aj} is the time at which queuing begins on route;
j is the beginning of the rush hour;
r_j is the departure rate from home along route.

Arrival rate   Is a function of the utility (or disutility) at time t of arriving at
that time which is, in turn, a function of the travel cost. Individual travel cost
is equal to the sum of any toll, plus the sum of the products of each of the
following: the shadow cost of time spent traveling and travel time; the shadow
cost of being late; and the time of a late departure. Travel cost at time t on
route j, or $C/(t)$, is described mathematically by the following (Arnott *et al.*,
1990):

$$C_j(t) = \alpha\, T_j(t) + \beta\, \text{(time early)} + \gamma\, \text{(time late)} \qquad (8.5)$$

where:
α is the shadow cost of total travel time;
β is the shadow cost of time when the driver leaves early;
γ is the shadow cost of time when the driver leaves late.

This equation is modified to show an individual's travel cost function when
they depart early:

$$C_j(t) = \alpha\, T_j(t) + \beta\, (t^* - t - T_j(t)) \qquad (8.6)$$

and when they depart late:

$$C_j(t) = \alpha\, T_j(t) + \gamma\, (t + T_j(t) - t^*) \qquad (8.7)$$

where:

t^* is the official work starting time;

t is the actual time of departure.

These cost functions can be aggregated to show trip cost (Arnott *et al.*, 1991):

$$C_j(t) = \alpha\, T_j(t) + \beta \max\, [0,\, t^* - (t + T_j(t))] + \gamma \max\, [0,\, t + T_j(t) - t^*\,] \quad (8.8)$$

Figure 8.5 illustrates a no-toll equilibrium with capacity sj and N number of drivers. T_n is the last time period in which a driver can depart and still get to work on time (t^*). After this point, the queue decreases, until point C, when queue length equals zero.

In the absence of a toll, trip cost can be calculated in the following way. The marginal cost of an additional traveler is derived from the marginal social cost curve, which is equal to:

$$2\delta(N/s_j) \quad (8.9)$$

where:

$\delta = (\beta\,\gamma)\,/\,(\beta + \gamma)$;

N is the number of commuters;

s_j is the capacity at the bottleneck on route j.

Using cost as a measure of disutility, arrival rates to the bottleneck are determined by a logic model, described as follow:

$$P(t) = \exp\,(-V(t))\,/\,\Sigma_t \exp\,(-V(t)) \quad (8.10)$$

$P(t)$ is the probability that an individual arrives at the bottleneck at time t, and $V(t)$ is the utility associated with arriving at the bottleneck at that time. To estimate the number of individuals who arrive at each time t, we use the following equation:

$$\text{ARR}(t) = N^*P(t) \quad (8.11)$$

where N is the total number of travelers.

The first bottleneck model as a policy analysis tool   The bottleneck model just described can be used to evaluate the effect of alternative policies relating to tolls and ITS. There are several tolling scenarios which might be used in a bottleneck model. The advantage of a same rate flat toll is that it is easier and cheaper to collect. It is more acceptable to the public since it is easier to

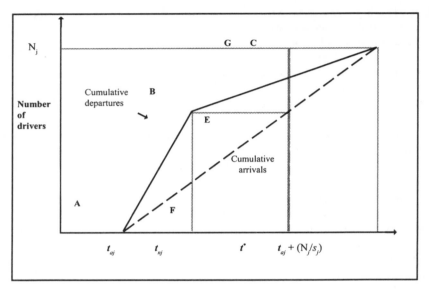

Source: Arnott *et al.* (1991)

Figure 8.5   Equilibrium departure rate and arrival distribution in equilibrium

understand. An optimal flat toll is equal to the expected marginal external costs with imperfect information. A variable toll is one that charges different rates, usually higher rates for peak period congestion. There are two types of variable tolls: step tolls and constant time-varying tolls. Step tolls charge a different pre-established rate at different times of the day. Fluctuating or time-varying tolls change in response to congestion and traffic volume. Fluctuating tolls are more expensive to collect, since the sums required vary, and have not been popular or widely used.

For time varying tolls, the fee should be equal to the marginal external cost price. This is equal to the difference between the marginal social cost of the trip and the portion of the average variable cost borne by the traveler (Small, 1994). An examination of step tolls versus time-varying tolls via electronic pricing can show the efficiency gains from the latter. The ratio of efficiency gains from the optimal step toll to a time-varying toll is equal to $(n/(n + 1))$. Thus, an optimal one step toll would yield half the efficiency gains of a time-varying toll, a two step, two-thirds, and so forth. An optimal time-varying toll is more efficient, and can be achieved by changing the time pattern of departures.

There are many arguments for congestion pricing. The seminal work by Vickrey (1967) uses economic arguments for tolling price schemes. According to Vickrey, public roads are no different than telephones, transit or any other service which uses peak period pricing for demand management. El Sanhouri and Bernstein (1994) offer three reasons why congestion pricing may influence the success of ITS. First, they argue that demand management strategies must be included in ITS measures, since the goal of these is to increase efficiency and capacity. Second, they believe that demand management strategies may improve the overall gains from ITS, since many ITS measures may not work with unpriced congestion. Finally, they argue that congestion pricing, which has been difficult to implement in the US, may be more politically acceptable as part of an ITS package.

Designing a system which deals with congestion by increasing road or highway efficiency can be done in one of two ways. First, tolls can be used to modify the driving behavior – specifically the departure times – of an individual who uses the road. Second, 'telematics' can be used: these include information technologies which provide additional information to drivers to modify their behavior, such as advance warning signs, route-guidance or pre-trip congestion messages. Unlike tolls, these have no direct cost to consumers, and therefore are politically more popular (Verhoef *et al.*, 1995). The two methods are to some degree complementary. Without advance warning of congestion, drivers are unlikely to alter their route choice or depart early to avoid paying higher tolls. Variable tolls assume that drivers have information about congestion, toll prices, and the value of their travel time. Thus, an assumption in these models is perfect information.

Arnott *et al.* (1991) examine the equilibrium effects of pre-trip information in the absence of tolls on travel costs. They find that without perfect information there may be no decrease in congestion on the whole, although information does benefit certain drivers. Drivers with access to pre- or en-route information will modify their behavior, knowing that others do not have the same information. Likewise, those who do not have access to trip information technologies alter their driving patterns based on expectations about the driving behavior of informed commuters. Information must be strategically provided to target populations in order to be useful.

A study by Verhoef *et al.* (1995) used five possible scenarios as alternatives for analyzing the interaction of congestion pricing and information provision: no tolling with imperfect information; no tolling with perfect information; flat (non-fluctuating) tolling with imperfect information; flat tolling with perfect information; and fluctuating tolling (which, as explained above, assumes perfect information). They assume an elastic demand (drivers are able to take alternate routes or change their travel times), and that information about congestion is purely "public" and available to all drivers. They use a stochastic travel cost function which is determined by the probability of congestion at network links, route capacity and transportation costs.

The role of information in reducing wait time at a bottleneck may be characterized in this model. The use of automated tolls that reduce waiting time at a queue have a value associated with them which can be derived using cost-benefit analysis. Information that is necessary to do this includes the cost of the automated toll and the value of the time it saves (how it changes the individual travel cost in the equation above). Another type of information which can be modeled is pre- or en-route information provided to drivers. These include variable message signs, warnings broadcast over the radio or television reports. The value of these may also be analyzed using this model, as described below.

Deciding on the type of pricing scheme and technology to be used involves the following steps. First, what are the goals of the system? In the case of the Woodrow Wilson Bridge, an attempt to simulate the management of peak hour congestion by altering departure times is made. Second, how can these goals be accomplished? The use of time-varying tolls may be an effective way to induce changes in departure times, as well as route choice assignments. Advanced information helps drivers make the choice which maximizes their utility. In this way, ITS can be used in conjunction with demand management strategies to alter driving behavior.

Policy simulation   Several policy scenarios relating to ITS and tolling are investigated using the general bottleneck model. These include: no toll; flat toll with manual collection; flat toll with electronic collection; step toll with manual collection; and step toll with electronic collection scenarios. In addition, three scenarios relating to the use of telematics are explored. Each alternative is compared in terms of average number of vehicles in the queue, maximum number of vehicles in the queue, duration of the queuing period, and consumer surplus or utility.

For each of these simulations, the following assumptions are made. Average daily traffic volume on the Woodrow Wilson Bridge is 160 000, while daily design capacity is only 84 000 vehicles. Values of the schedule delay parameters are assumed to be $t^* = 8{:}30$ a.m., $t_0 = 5{:}00$ a.m., $t = 12{:}00$ a.m., $\alpha = \$6/\text{hr}$, $\beta = \$3/\text{hr}$, $\gamma = \$9/\text{hr}$. The departure period is divided into one-minute intervals. Demand for travel on the Woodrow Wilson Bridge is assumed to be relatively inelastic and commuters are assumed to have a medium amount of information concerning travel conditions on the bridge. Information is considered "public": either every driver or no driver has access to transportation and tolling information. Information can be provided through: television or radio traffic announcements; variable or advance warning message signs; newspaper announcements of tolling price or price changes; or it can be the prod-

Table 8.7   Summary of tolling alternatives

Scenarios	No toll	Flat toll (manual collection)	Flat toll (electronic collection)	Step toll (manual collection)	Step toll (electronic collection)
Average queue length	234.00	324.00	234.00	0	0
Maximum queue length	563.00	695.00	563.00	0	0
Duration of queue (hrs)	6.10	6.85	6.10	0	0
Total consumer surplus	123.36	44.45	45.38	6.19	7.85
Average consumer surplus	0.29	0.11	0.11	0.02	0.02

uct of travel experience on the bridge. Simulations are performed only for inbound a.m. traffic on the bridge, although the model could just as easily be applied to outbound traffic as well.

Table 8.7 summarizes the results of the analysis of the five tolling alternatives. A number of observations can be made. First, in terms of reducing average queue length, maximum queue length, and duration of the queuing period, step tolls appear to be the most effective. At the same time, total and average consumer surplus appears to be much lower with this type of policy than with a flat or no toll scenario. Perhaps this could be explained by the fact that there is a relatively high disutility associated with schedule delay and tolling. Second, in the case of either flat or step tolling, electronic collection provides commuters with a higher level of consumer surplus and a lower amount of delay. Manual toll collection results in a reduction in capacity, which increases delay, as illustrated in Figure 8.6. Manual toll collection also requires that a higher step toll be charged to commuters, as shown in Figure 8.4.

Table 8.8 summarizes the effects of providing commuters with varying levels of information concerning delay and tolling. "High" represents a situation in which all commuters are provided with the same type and amount of information. In terms of reducing average queue length, maximum queue length, and duration of queue, a "high" use of telematics appears to be the most effective. On the other hand, it provides commuters with the least amount of utility or consumer surplus.

And what about the impact of a time-varying toll? One can hypothesize about this scenario. The type of toll which is implemented affects whether or

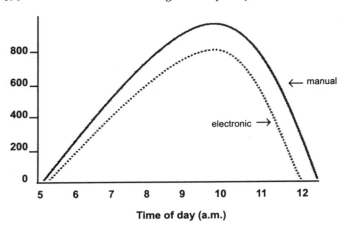

Figure 8.6   Method of toll collection and delay

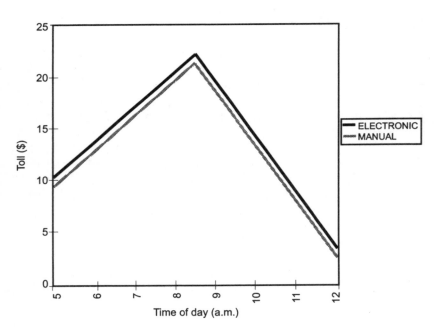

Figure 8.7   Optimal step tolls by method of collection

Table 8.8   Summary of telematics alternatives

	Low	Medium	High
Average queue length	483.00	234.00	134.00
Maximum queue length	1145.00	563.00	340.00
Duration of queue (in hours)	6.20	6.10	5.78
Total consumer surplus	184.85	123.36	37.54
Average consumer surplus	0.44	0.29	0.09

not congestion will be reduced in the long term. Drivers who face a fixed toll know what the rate is going to be, the value of their time, and the time it takes to travel to work. They will leave at a time estimated by the travel cost. Drivers can also make informed decisions with variable tolls that change at fixed times. However, with an optimal flexible toll, the amount charged varies continuously based on congestion patterns. Drivers are unable to make informed decisions without knowing the toll rate that they will encounter. Thus, similar to the conclusions of Arnott *et al.* (1991), information about this kind of toll can actually exacerbate congestion problems.

The second bottleneck model
In the first model, the rate of flow of cars (cars per unit time that pass through the WWB) is considered to be a constant value. However, this is not realistic because bottleneck capacity has some relation to the length of the queue.[1] While the queue is small, the rate of flow S is close to the capacity. But as the queue gets large, it will have an effect on the rate of flow. In other words, the length of the queue will affect the speed of the cars as they approach the bottleneck and therefore affect the rate of flow S. Although a minute-by-minute rate of flow is not available for the WWB, the rate of flow in each hour shows a change in the flow during the rush hours (when queuing is present), proving that the rate of flow is not always the same as the capacity rate, and is dependent on the length of queue (USDOT, 1996b). A look at the peak hour speeds on the Woodrow Wilson Bridge shows that the speed on the bridge will decrease to a level between 21 and 30 m.p.h. (USDOT, 1996b). This is less than half the value of the normal 55 m.p.h. speed of non-peak hours.

Model specification and calibration   Therefore the rate of flow fluctuates from a high point of the bottleneck's capacity of S_{max} to a minimum flow of S_{min}, and is dependent on the length of the queue:

$$S = S_{min} + (S_{max} - S_{min}) / [1 + (e^{-a + bQ})] \qquad (8.12)$$

This equation means that with no queue present, or with small queues, the exponential term will go to zero (since e^{-a} is very small), and the rate of flow will reach the maximum:

$$S \circledR S_{min} + (S_{max} - S_{min}) \quad (as \ Q \circledR 0) \tag{8.13}$$

therefore:

$$S \circledR S_{max} \ (as \ Q \circledR 0) \tag{8.14}$$

The equation above is the special case where the rate of flow equals the capacity of the bottleneck, therefore:

$$S_{max} = the \ capacity \ of \ the \ bottleneck \tag{8.15}$$

As the queue gets larger the term bQ will become more significant and with very large queues the exponential term in the equation will approach infinity:

$$S \circledR S_{min} \quad (as \ Q \ gets \ very \ large) \tag{8.16}$$

which means that with very large queues, the rate of flow will reach a minimum.

Figure 8.8 shows this fact to some degree. The data shown are the hourly rate of flow and queue length. Since maximum rush hour happens in a period of one to two hours, to accurately show the effect of the queue on the rate of

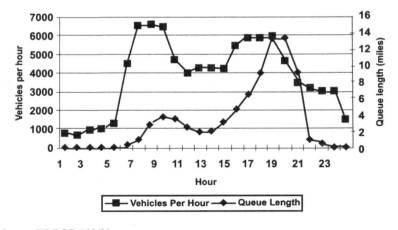

Source: US DOT (1996b)

Figure 8.8   Hourly rate of flow and queue length (mile)

flow, minute-by-minute queuing data along with rate flow per minute is needed. Nevertheless, even with these data, a lag time of about an hour is seen between the queue and the rate of flow; and the rate of flow shows a decreasing trend as the queue reaches a maximum.

To present a simplified sample of this model, we let S_{max} be the maximum hourly rate of flow 6500. S_{min} is taken to be the minimum rate of flow at maximum queue. The longest queue built up on the Woodrow Wilson bridge is 13.5 miles.

Relation to ITS Currently, the rate of flow of cars is only known for every hour. Through the use of simple ITS technologies such as counters, we can detect and record the rate of flow of vehicles for every minute on the Woodrow Wilson Bridge. These data would provide both the maximum and the minimum rate of flow of vehicles through the bottleneck. The maximum rate S_{max} would be used for the capacity of the bottleneck. The minimum rate is the exact value of S_{min}. Of course the data for an ample time period should be taken to account for non-receiving incidents and traffic jams not caused by queuing alone.

ITS can also provide tools in order to measure the length of the queue at continuous intervals. This will give an estimate of the value for Q. Having the value for Q, for different rates of flow, solving for the best combination of *a* and *b* is possible.

Through the use of ITS, it would be possible to produce a continuous chart of the queuing behind the bottleneck and its corresponding rate of flow through the bottleneck. One benefit of such information could be better incident and accident management. Through automatic monitoring of the length of queue, the rate of flow could be calculated. If this rate is less than the average rate of flow for such a queue, another incident – other than normal queuing traffic – is causing the problem; subsequently, authorities could be notified. Additionally, automatic monitoring of the queue (and the rate of flow) with the use of this model will show the effect of any change in the operationality of the bottleneck (for example, construction, and new HOV lanes); and the model will give an instant diagnosis of the effect of such changes.

The model introduced shows how simple ITS technologies can provide tools for faster detection of incidents and immediate response to traffic conditions. It can also replace alternative higher priced surveillance systems. This model could also be utilized for other bottlenecks in the area for a comprehensive incident management system.

CONCLUSIONS

This case study began with the assumption that in evaluating ITS benefits, traditional methods have some shortcomings. Based on the information gathered from the results of the methodologies used, the evaluation checklist, which summarized the effects of ITS technology on the Woodrow Wilson Bridge regarding a variety of categories, has been given. Historical background and cultural tradition also influenced this summary. In order to look at all the costs and benefits of ITS, and to address complex interaction effects, three supplemental methodologies were applied.

The first methodology was the PROSCAL analysis of the preferences of ITS benefits. The survey was an aid in establishing collective expert preferences of ITS priorities. PROSCAL – which proved to be an alternative to conventional multi-criteria utility models in evaluating ITS decision making – highlighted the connection between the decision and the variance of individual judgments due to information, experience, and the psychological background of decision makers, as well as the individuals affected by those decisions. What this information has provided is the perception of both ITS experts and potential ITS users; this in turn could set the priority of policy makers for devoting resources to such institutional issues as privacy, social equity, mobility, access, and information gathering. Although the result of the survey was not conclusive in all respects, it was possible to differentiate between the top four and the remaining five preferences. The top four preferences of ITS benefits were: improved safety; better traffic flow; better environmental quality; and better travel information.

The second supplemental methodology, meta-analysis, was used as a tool to identify evaluation lessons on congestion pricing in the published literature on that subject. It was found that meta-analysis went beyond other forms of analysis in providing information on institutional, political, social and economic impacts of ITS. Although in this analysis data from studies related to the WWB situation were used, there was insufficient available data. As suggested in the assessment of the analysis results, the deployment of ITS technologies can certainly remedy this, and can also help to obtain better information in a more efficient manner.

The next methodology studied was the bottleneck model. The first application of the bottleneck model demonstrated the magnitude of the potential efficiency gains that congestion tolls can achieve by changing the time pattern of departures without significant reductions in the number of trips. The model also showed its value in evaluating the effects of alternative policies related to tolling and ITS.

The second application of the model provided a tool for finding an alternative solution for incident and accident management. It has also helped identify a method in which lower-cost ITS technologies can be substituted for alternative higher-priced surveillance systems. In this case, it was proposed that electronic counters be used to measure the rate of flow and the length of the queue. Using this information and the model provided, unusual traffic jams can be detected without the use of video-monitoring, which is the higher-priced alternative. Also with the use of this model, an instant diagnosis of any change in the operation of the bottleneck can be provided.

The three methodologies mentioned have shed more light on the benefits of ITS as they apply to the Woodrow Wilson Bridge. They have shown the preferences of decision makers and users of the bridge for different ITS benefits and have introduced new methods to be used in understanding these benefits and their impacts. But above all, what these methodologies and new approaches have provided are new windows of opportunity that are not available without the assistance of ITS technologies. The bottleneck model is a perfect example of what ITS provides in new ways of solving transportation problems – an alternative to building new roads and infrastructure. What ITS provides is complex technologies that are tools in understanding and solving tomorrow's complex transportation needs. Solutions such as the bottleneck model and the congestion pricing model used in this study are examples of these methods. Implementation of such models without the use of ITS is nearly impossible, and most probably not cost-efficient. Therefore, along with the question of how to measure the benefits of ITS, an attempt has been made to answer the question of what ITS technologies are the right technology options, and what ITS benefits are the right benefits to provide solutions for the complex transportation problems of tomorrow.

APPENDIX 1: CHECKLIST: USE OF ITS ON THE WOODROW WILSON BRIDGE

Category	Goal	Indicator	Effect	Comment
Congestion and delay		Peak-hour VMT	Increase	Significantly
		Vehicle-hours of delay	Decrease	
		Volume/capacity ratio		
		Peak/non-peak travel times	Decrease	
		Average speed (e.g. PMT/P-Hr)	Increase	
		Average length of queue	Decrease	
Capacity and mobility		Ease of personal mobility	Increase	Significantly
		System accessibility	Increase	
		Network or connectivity implications	Increase	
		Passenger ridership	Increase	
		Daily VMT	Increase	
Safety and Accidents		Deaths	Decrease	
		Injuries	Decrease	
		Property damage	Decrease	
Cost relative to other options		Construction	Decrease	
		Operations	Decrease	
		Maintenance	Decrease	
Financial and fiscal outcomes		Revenue enhancement	Increase	Tolls generate revenue
		Budgetary sufficiency	Improved	Tolls fund bridge construction

Category	Goal	Indicator	Effect	Comment
		Inter-governmental transfers	None	
		Private profits	Not sure	
Environmental outcomes		Total emissions	Uncertain	
		Ground-level ozone	Not sure	
		Energy consumption	Decrease	
		Hazardous materials impact	No effect	
		Soil erosion	No effect	
		Wetlands and habitat	No effect	
Economic outcomes		Gross regional product	Increase	
		Employment	Increase	
		Per capita income	Increase	
		Land values	Increase	
		Regional competitiveness	Increase	
Equity outcomes		Affordability	Decrease	
		Regional access disparity	Worse	
		Handicap access	Not sure	
Institutional outcomes		Flexibility, adapts	Not sure	
		Regional significance	Yes	
		Synergy	Uncertain	
		Intergovernmental friction/cooperation	Yes	
		Community involvement and preferences	Yes	
		Ownership changes	Yes	
		Public–private ptnrs	Yes	
		Liability	Yes	

Category	Goal	Indicator	Effect	Comment
		Privacy	Yes	
Spatial outcomes		Land use implications	Yes	
		Land values	Increase	
		Residential proximity issues		
		Infrastructure design or geometry impact	Yes	
Demonstration outcomes		Transferability	Yes	
		Value as a model	Yes	
		Suitability		

APPENDIX 2: ITS TECHNOLOGIES

Intelligent Transportation Systems include four categories:

1. *ATMS (Advanced Traffic Management Systems)* These include technologies which help to utilize highway capacity and reduce congestion, accident rates and trip times. Technologies we suggest for evaluation on the bridge may include ramp metering and incident management systems.
2. *AVCS (Automated Vehicle Control Systems)* These technologies are designed to help drivers on the highway. We included automated tolls.
3. *ATIS (Advanced Traffic Information Systems)* These provide information on congestion, navigation and location, traffic conditions, and alternative routes. ATIS systems we included are traffic diversion systems, alternate route and advanced warning (information broadcasting) systems, and systems which provide information to safety officials about conditions on the bridge.
4. *CVO (Commercial Vehicle Operations)* These technologies help commercial vehicles operate more efficiently. Rolling weight stations and real time sensor weighing systems are examples of high technology CVO technologies.

The five relevant purposes of ITS that are important for this study are congestion management, flow improvement, safety and incidence management, cost recovery and information generation.

APPENDIX 3: INDIVIDUAL LEVEL ANALYSIS – EUCLIDEN DISTANCES BY BENEFITS

Respondents	Benefits								
	1	2	3	4	5	6	7	8	9
1	0.87	0.81	1.00	0.75	2.20	2.43	1.93	0.56	1.43
2	0.87	0.81	1.00	0.75	2.20	2.43	1.93	0.56	1.43
3	1.33	1.28	1.47	1.22	2.66	2.89	1.47	0.098	0.97
4	1.20	1.15	1.33	1.09	2.53	2.76	1.60	0.23	1.10
5	2.58	2.64	2.45	2.70	1.25	1.02	5.38	4.01	4.89
6	1.26	1.20	1.39	1.14	2.59	2.82	1.54	0.17	1.04
7	0.65	0.60	0.79	0.54	1.99	2.21	2.14	0.77	1.64
8	0.71	0.66	0.85	0.60	2.04	2.27	2.09	0.72	0.16
9	0.87	0.81	1.00	0.75	2.20	2.43	1.93	0.56	1.43
10	3.13	3.08	3.27	3.02	4.46	4.69	0.33	1.70	0.83
11	1.01	1.06	0.87	1.12	0.32	0.55	3.81	2.44	3.31
12	0.87	0.81	1.00	0.75	2.20	2.43	1.93	0.56	1.43
13	1.53	1.58	1.39	1.64	0.20	0.03	4.33	2.96	3.83
14	0.30	0.25	0.44	0.19	1.63	1.86	2.50	1.13	2.00
15	1.05	0.99	1.18	0.94	2.38	2.61	1.75	0.38	1.25
16	1.81	1.76	1.95	1.70	3.15	3.37	0.98	0.38	0.49
17	0.24	0.18	0.37	0.12	1.57	1.80	2.56	1.19	2.06
18	0.87	0.81	1.00	0.75	2.20	2.43	1.93	0.56	1.43

| | Benefits | | | | | | | | |
Respondents	1	2	3	4	5	6	7	8	9
19	0.75	0.70	0.89	0.64	2.09	2.31	2.04	0.67	1.55
20	2.59	2.65	2.46	2.70	1.26	1.03	5.39	4.02	4.89
21	1.20	1.15	1.34	1.09	2.54	2.77	1.59	0.22	1.09
22	0.07	0.02	0.21	0.04	1.40	1.63	2.73	1.36	2.23
23	0.87	0.81	1.00	0.75	2.20	2.43	1.93	0.56	1.43
24	0.55	0.49	0.68	0.44	1.88	2.11	2.25	0.88	1.75
25	0.87	0.81	1.00	0.75	2.20	2.43	1.93	0.56	1.43
26	0.63	0.58	0.77	0.52	1.97	2.20	2.16	0.79	1.67
27	1.37	1.31	1.50	1.25	2.70	2.93	1.43	0.06	0.93
28	0.87	0.81	1.00	0.75	2.20	2.43	1.93	0.56	1.43
29	0.62	0.56	0.75	0.50	1.95	2.18	2.18	0.81	1.68
30	4.50	4.45	4.64	4.39	5.84	6.06	1.70	3.07	2.20
31	1.45	1.39	1.58	1.33	2.78	3.01	1.35	0.02	0.85
32	2.43	2.48	2.29	2.54	1.10	0.87	5.23	3.86	4.73

APPENDIX 4: TABLES OF SURVEY RESULTS

Table 8A.1   Number of years of transportation management experience

Experience	Frequency
No experience	2
1 to 5 years	7
6 to 10 years	7
11 to 15 years	7
16 years or more	12
Total	35

Table 8A.2   Number of years directly involved in ITS activities

Experience	Frequency
No experience	17
1 to 5 years	12
6 to 10 years	3
11 to 15 years	1
16 years or more	2
Total	35

Table 8A.3   Employment activity

Activity	Frequency
Urban planning	0
Transportation planning & analysis	24
ITS/IVHS planning and/or management	3
Law enforcement/fire & rescue	2
Other	6
Total	35

Table 8A.4   Woodrow Wilson Bridge planners/researchers

Planners/researchers	Frequency
Yes	10
No	25
Total	35

Table 8A.5   Woodrow Wilson Bridge commuters

Commuters	Frequency
Yes	3
No	32
Total	35

Table 8A.6   Work location

Location	Frequency
Fairfax County	5
Prince William County	1
Loudoun County	2
Arlington County	1
Alexandria	1
District of Columbia	14
Montgomery County	3
Prince George's County	4
Other	4
Total	35

Table 8A.7   Organization affiliation

Affiliation	Frequency
Federal or state department of transportation	13
Council of Governments	5
Law enforcement/fire & rescue	2
Commercial vehicle management	0
Private transportation contractor/consultant	1
Local government	14
Other	0
Total	35

NOTE

1. For many reasons this argument was not considered in the first model. First, no mention of this effect was made in the literature found on bottleneck effects. The second reason was to keep an already complex model from being overly convoluted. But above all, the reason for exclusion of this argument in the first model was lack of enough data to provide an exact model of the effect that the length of queue has on the rate of flow. The model is however mentioned here because of its significance to ITS.

REFERENCES

Arnott, Richard, Palma, Andre de and Lindsey, Robin (1990), "Economics of a bottleneck", *Journal of Urban Economics* **27**, 111–30.

Arnott, Richard, Palma, Andre de and Lindsey, Robin (1991), "Does providing information to drivers reduce traffic congestion?", *Transportation Research (A)* **25A**(5), 309–18.

Bernstein, David and Muller, Jerome (1993), "Understanding the competing short-run objectives of peak period road pricing", *Transportation Research Record* **1395**, 122–8.

Bowen,W. and Haynes, K. (1994), "Environmental priorities and individual differences: metropolitan Cleveland", *Environmental Professional* **16**, 304–13.

Button, Kenneth (1994), "What can meta-analysis tell us about the implications of transport?", *Regional Studies* **29**(6), 507–17.

Combs, C.H. (1963), *A Theory of Data*, New York: John Wiley.

Daniel, Joseph. (1993), "Peak-load-congestion pricing of hub airport operations with endogenous scheduling and traffic-flow adjustments at Minneapolis-St. Paul airport", *Transportation Research Record* **1298**, 1–13.

De Corla-Souza, Patrick (1994), "Applying the cashing out approach to congestion pricing", *Transportation Research Record* **1450**, 34–7.

El Sanhouri, Ibrahim and Bernstein, David (1994), "Integrating driver information and congestion pricing systems", *Transportation Research Record* **1450**, 44–50.

Frick, Karen T., Heminger, Steve and Dittmar, Hank (1996), "Bay Bridge congestion pricing project: lessons learned to date", paper presented at the 75th Annual Meeting of Transportation Research Board, January 7–11.

Gifford, Jonathan L. and Talkington, Scott W. (1994), "Demand elasticity under time varying prices: a case study of the Golden Gate Bridge", in *Proceedings of the IVHS America 1994 Annual Meeting*, Atlanta, Georgia, pp. 503–7.

Gomez-Ibanez, Jose A. (1992), "The political economy of highway tolls and congestion pricing", *Transportation Quarterly* **46**(3), 343–60.

Goodwin, P.B. (1992), "A review of new demand elasticities with special reference to short and long run effects of price changes", *Journal of Transport Economics and Policy*, May, 155–69.

Haynes, Kingsley E., Arieira, Carlos, Burhans, Sarah and Pandit, Nitin (1995), "Fundamentals of infrastructure financing with respect to ITS", *Built Environment* **21**(4), 246–54.

Haynes, Kingsley E. and Li Qiangsheng (1993), "Policy analysis and uncertainty: lessons from the IVHS transportation development process", *Computers, Environment and Urban Systems* **17**, 1–14.

Heffner, R.A. (1958), *Extensions of the Law of Comparative Judgement to Discriminable and Multidimensionable Stimuli*, Ann Arbor, MI: University of Michigan.

Higgins, Thomas J. (1994), "Congestion pricing: implementation considerations", *Transportation Quarterly* **4**(3), 287–98.

Hirschman, Ira, McKnight, Claire, Pucher, John, Paaswell, Robert E. and Berechman, Joseph (1995), "Bridge and tunnel toll elasticities in New York", *Transportation* **22**, 97–113.

JHK & Associates (1995), *Woodrow Wilson Bridge Improvement Study: Executive Summary*.

Levinson, Herbert S. (1994), "Freeway congestion pricing: another look", *Transportation Quarterly* **1450**, 8–12.

Mackay, D.B. and Zinnes, J.L. (1988), "Probabilistic multidimensional scaling of spatial preferences", in R.G. Golledge and H. Timmermans (eds), *Behavioral Modeling in Geography and Planning*, New York: Methuen pp. 198–222.

McCarthy, Patrick S. and Tay, Richard (1944), "Pricing road congestion: recent evidence from Singapore", *Policy Studies Journal* **21**(2), 296–308.

McDonald, John F. (1995), "Urban highway congestion: an analysis of second-best tolls", *Transportation* **22**, 353–69.

McMullen, B. Starr (1993), "Congestion pricing and demand management: a discussion of the issues", *Policy Studies Journal* **21**(2), 285–95.

Metropolitan Washington Council of Governments (1995), *Woodrow Wilson Bridge Improvement Study, Regional Travel Demand Analysis, Executive Summary*.

Nijkamp, Peter (1996), "More studies, more insights? Reassessing spatial and environmental research by means of meta-analysis and rough set analysis", paper presented at the Fifth World Regional Science Association International Conference, Tokyo, May 7–10.

Odeck, James and Skjeseth, Trond (1995), "Assessing Norwegian toll roads", *Transportation Quarterly* **49**(2), 89–98.

Orski, C. Kenneth (1992), "Congestion pricing: promise and limitations", *Transportation Quarterly* **46**(2), 157–67.

Oum, Tae Hoon, Waters, W.G. II and Jong-Say Yong (1992), "Concepts of price elasticities of transport demand and recent empirical estimates: an interpretative survey", *Journal of Transport Economics and Policy*, May, 139–54.

Perera, Max H. (1992), "Framework for classifying and evaluating economic impacts caused by a transportation environment", *Transportation Research Record* **1274**, 41–52.

Polydoropoulou, Amalia, Ben-Akiva, Moshe and Kaysi, Isam (1994), "Influence of traffic information on drivers' route choice behavior", *Transportation Research Record* **1453**, 56–65.

Poole, Robert W. Jr and Yuzo Sugimoto (1995), "Congestion relief toll tunnels", *Transportation* **22**, 327–51.

Ross, R.T. (1934), "Optimum orders for the presentation of pairs in the method of paired comparisons", *Journal of Educational Psychology* **25**(5), 375–82.

Small, Kenneth A. (1993), "Urban traffic congestion: a new approach to the Gordian knot", *Brookings Review* **11**(2), 6–11.

Small, Kenneth (1994), "Urban transportation economics", in *Fundamentals of Pure and Applied Economics*, Boston, MA: Harwood Academic Publishers.

Stough, Roger R., Maggio, Mark E. and Dengjian Jin (1995), "Evaluation of regional benefits of an intelligent transportation system (ITS) application: a case study of electronic toll collection", presented at the Western Regional Science Association Meeting, February 20.

Thurstone, L.L. (1927), "A law of comparative judgement", *Psychological Review* **34**, 273–86.

Transportation Research Board (1985), *Highway Capacity Manual*, Special Report 209.

US DOT (1995), *Assessment of ITS Benefits: Early Results*. Prepared under Contract by Mitre, Sponsored by the Federal Highway Administration. August 1995, Washington, DC, 3-23.

US DOT (1996a), *Advanced Public Transportation Systems: The State of the Art, Update '96*. US Department of Transportation. Federal Transit Administration.

US DOT (1996b), *Woodrow Wilson Bridge Improvement Study*, Transportation Technical Report 3-23, prepared by US Department of Transportation Federal Highway Administration, Region 3 and Virginia Department of Transportation, Maryland Department of Transportation, State Highway Administration, and District of Columbia Department of Public Works, January.

Verhoef, Erik, Emmerink, Richard, Nijkamp, Peter and Rietveld, Piet (1995), "The economics of information provision and road pricing with stochastic road congestion", paper presented to the 42nd North American Meeting of the Regional Science Association International, Special Session on Road Pricing and Traffic Congestion, Cincinnati, Nov. 9–12.

Vickrey, William S. (1967), "Pricing and resource allocation in transportation and public utilities: pricing in urban and suburban transport", *American Economic Association* 452–65.

Williams, Alan W. (1995), "Should the user pay? Lessons from Anglo-Australian history", *Transportation* **22**, 115–34.

Wilson, P. (1988), "Welfare effects of congestion pricing in Singapore", *Transportation* **15**(3), 191–210.

Woodrow Wilson Bridge Study and Design Center (1995), *Woodrow Wilson Bridge Improvement Study*, Federal Project I-95-2 (233) 182. State Project 0095-100-104, PE 104.

Zinnes, J.L. and Mackay, D.B. (1987), "Probabilistic multi-dimensional analysis of preference ratio judgements", *Communications and Cognition* **20**(1), 17–24 and Appendix.

9. Incident Management and Intelligent Transportation Systems Technology: Estimating Benefits for Northern Virginia

Gerard Maas, Mark Maggio, Hadi Shafie and Roger R. Stough

INTRODUCTION

Metropolitan areas across the US are faced with increasing congestion (Institute of Transportation Engineers, 1996). This chapter deals with estimating the benefits of the deployment of Intelligent Transportation Systems (ITS) technology for the purpose of aiding incident management (IM) and reducing non-recurrent congestion from freeway incidents. Non-recurrent congestion from traffic accidents and other incidents is a major component of the overall congestion problem and is being addressed through IM programs. In order to maximize benefits from the deployment of new technologies, it is necessary that IM be augmented with supporting ITS systems. However, building a mandate and obtaining sufficient resources to achieve this is often hampered by an inability to satisfactorily quantify the expected benefits from IM programs or the application of ITS technology in such programs. This chapter examines this benefit estimation problem, develops an estimation methodology, and applies it in a Northern Virginia context.

The first section discusses congestion problems, and ITS and IM as responses to these problems. The benefit estimation issue is then assessed. In the second section, these topics are discussed in the context of the Northern Virginia region. An approach to the estimation of the benefits of ITS in IM in Northern Virginia is described in the third section. The results of several benefit estimations for the Northern Virginia region are presented in the fourth section. Conclusions are presented as the final section of the chapter (US Department of Transportation, 1987).

STRATEGIES FOR DEALING WITH URBAN CONGESTION

Each year, traffic congestion results in millions of hours of vehicle delay, with associated increases in fuel consumption, environmental pollution, and productivity losses. In monetary terms, the cost of traffic congestion to the US economy has increased in recent decades and is expected to rise to $88 billion per year by 2005 (Institute of Transportation Engineers, 1996). By that date, urban freeway delays are expected to reach 4 billion hours annually if present conditions remain unchanged, an increase of 436 percent relative to 1985 conditions (US Department of Transportation, 1987). Congestion poses a particular challenge to the viability of urban centers, metropolitan areas, and core economic regions.

Recurrent congestion is due to high levels of traffic demand during peak travel hours which outstrip the available roadway capacity. Non-recurrent congestion results from traffic accidents and other incidents that cause a temporary and unanticipated reduction in road capacity. Traffic incidents can compound the problem of recurrent congestion, especially when they occur close to or during peak hours. Although incidents are essentially random events, peak hour traffic conditions may contribute to an increased occurrence of certain types of traffic incidents. Presently, nearly 60 percent of traffic delays can be classified as non-recurrent (Cambridge Systematics Inc., 1990). The Federal Highway Administration (FWHA) estimates that by 2005 congestion due to all types of traffic incidents will constitute 70 percent of all congestion, with a cost to road users of $35 billion (Mannering *et al.*, 1992).

Various approaches to dealing with recurrent congestion have been proposed and are being implemented. Capacity expansion can alleviate congestion problems but may not be a solution that is sustainable in the long run. In many localities, there are fiscal and spatial constraints on capacity expansion, and what new capacity can be created tends to be absorbed quickly as latent demand is released. Change on the demand side through traffic demand management (TDM) is essential and will increasingly be necessary. However, influencing travel and transportation related attitudes and behavior, for example, through pricing schemes, is a highly complex long-term endeavor that involves many factors generating high uncertainty, poor predictability and impacting sensitive equity considerations. Efforts to reduce recurrent congestion typically involve a mix of roadway capacity expansion, travel demand management strategies, and promotion of transportation modes other than the automobile. In view of the limitations of capacity expansion as a structural solution and the complexities of travel demand management as a strategic solution, assuring optimal use of existing transportation infrastructure in general, and roadway capacity in particular, has become an increasingly important issue.

Intelligent Transportation Systems

It is in the context of optimization of the usage of existing transportation infrastructure that the new and emerging technologies collectively known as Intelligent Transportation Systems find their application. ITSs incorporate a smoothing or rationalizing process intended to help existing supply of transportation infrastructure better meet current and future transportation demand. By employing advanced technologies, the infrastructure can be made more efficient, effective and safer while precluding increased land-use and environmental pollution problems. In addition, ITSs serve as a policy facilitator through which public policy can more effectively encourage energy saving and discourage road travel (Haynes *et al.*, 1994). Much ITS research has been technology oriented and pointed towards the improvement of the "engineering subsystem" (Padgett, 1996). However, ITSs also comprise information and human subsystems. Behnke (1996, p. 92) notes, for example, that "The transportation-related problems of most American communities are largely the result of not having the information system that will permit them to utilize their transportation resources effectively." ITS will impose new demands for larger amounts of real-time data not currently collected while at the same time greatly facilitating the collection and management of those data. Nevertheless, no matter how sound the technologies or how reliable the data, in the end it is the behavior of people who use, influence, or are influenced by, transportation technology that will ultimately determine how successfully ITSs achieve their intended objectives (Baird, 1996). They have economic, social, political, institutional-legal, environmental, and physiological impacts that affect success over and above the "engineering success". The paradigm shift in transportation technology and vision, as motivated by the emergence of advanced transportation components like ITS, may require a change in deployment strategies and may require new evaluation methods (Jin and Stough, 1996). ITS exhibit special inter-sectoral and inter-temporal effects, which make it imperative that a non-traditional, total cost and benefit approach to evaluation be adopted (Stough *et al.*, 1994).

Incident Management

Optimization of the usage of existing transportation infrastructure also provides the context for examining incident management here. Because traffic accidents and other incidents are a major source of congestion, a growing number of states and metropolitan areas are initiating or expanding IM programs. The specific purpose of these programs is to mitigate non-recurrent congestion and delays due to incidents. The goal of IM is to detect, verify, respond to and clear temporary obstructions on roadways in an expedient manner, so that normal traffic flow may be restored as rapidly as possible. A combination of strategies and systems is typically employed, including inter-jurisdictional and inter-agency emergency response planning and coordination, freeway service patrols, and enhanced traffic surveillance. Most IM efforts, however, comprise a patchwork of agency interactions and response activities that are combined in an ad hoc manner. Very few IM programs are strategically organized, with dedicated funding and clear lines of authority. IM programs lack visibility to the general public and are often perceived as ineffective (Cambridge Systematics Inc., 1997). The single most important step in the development and sustenance of an IM program is to build a mandate for it. Any mandate-building strategy involves efforts to raise the awareness and interest of public and private stakeholders in IM. Furthermore, raising the level of awareness to the point where stakeholders are ready to make organizational changes and commit resources to the program is essential. The provision of accurate information concerning the benefits of IM programs helps raise awareness by demonstrating that these are a cost-effective means of dealing with non-recurrent congestion (ibid.).

As it is the objective of the US Department of Transportation (US DOT) to see ITS technology deployed in 75 major metropolitan areas over the next decade, the linkage of IM to ITS deployment may provide additional momentum (Cambridge Systems Inc., 1997). Generating accurate estimates of benefits (Mannering *et al.*, 1992) from current or future ITS applications in the specific context of IM may help facilitate the development of this linkage. Although the number of IM programs is growing, attempts to strengthen funding for IM or to obtain funding for the deployment of ITS technology in IM are often stymied. Funding for IM is limited and subject to intense competition from more traditional highway projects (Cambridge Systematics Inc., 1997). State, local and municipal government agencies report difficulties in preparing budget justifications and cost-benefit analyses of new IM approaches and technologies.

Agency managers and policy makers are often unable to satisfactorily quantify the benefits of new ITS technology applications in IM, or even of existing IM efforts. Many IM programs are regional in scope or focus on an extended corridor (such as I-95) and thus involve "spill-over" benefits and costs outside the immediate application area. Spill-over effects are often underestimated and compound the difficulty of quantifying benefits. In order to create and sustain effective IM programs, it is essential to develop an accurate means of estimating the benefits of IM in general and related ITS applications in particular.

THE NORTHERN VIRGINIA CONTEXT

The traffic situation in Northern Virginia, the Virginia portion of the National Capital region, is very much determined by its location along I-95 and its proximity to the Washington, DC metropolitan region. The National Capital region has the distinction of being among the three most congested metropolitan areas in the US as measured by traffic delays. It also has the highest aggregate cost of delay in the nation. Projections for the National Capital region are similar to those for the nation, if not more pronounced. Travel in the region is increasing at an estimated rate of 4 percent per year (Lo *et al.*, 1994) and vehicle miles of travel (VMT) in 2020 is expected to be twice that of 1988 (Mobility 2000, 1990a). Trip lengths are expected to increase while home-to-work car occupancy is expected to decrease (Greater Washington Transportation Planning Board, 1990a). By 2010, vehicle demand will far outstrip highway capacity, not only on the Capital Beltway and major highways but also throughout the remainder of the area's road system. Between 1989 and 2009, a five-fold increase in freeway congestion is expected for the region (Mobility 2000, 1990b). Although incident rates are decreasing relative to the amount of travel, the consequences of these accidents and other incidents on the region's congested freeways are becoming more and more severe due to the increasingly heavy loading of the area's road system. If present trends continue, congestion and costly delays due to incidents will become substantially more prevalent than they already are today (Greater Washington Board of Trade, 1997a and 1997b).

A formal incident management program has existed in Virginia since the early 1980s. Funding for incident management has come from federal, state and local government sources as well as from motorist reimbursements. A variety of detection and verification techniques are currently employed. These include: routine police patrols; courtesy patrols (by police, public sector agencies, and private sector entities); a cellular phone emergency number; commercial traffic reports; automatic detectors; and closed circuit television. A range of response and clearance techniques have been established, including a formal response procedure, a quick clearance policy, and the use of Traffic Operations Centers, Traffic Management Teams and Incident Management Teams. There are agency-owned tow trucks as well as towing rotation agreements and towing contracts with private service providers. Additional response and clearance techniques are planned, such as an incident command system, an emergency response plan, HAZMAT response, and personnel resource lists. Recovery and information techniques used in Northern Virginia include traffic management, highway advisory radio (HAR), variable message signs (VMS), an 800 number for motorist information, a construction news service, and several media partnerships (Cambridge Systematics Inc., 1997).

Incident management in Northern Virginia is no less difficult then elsewhere in the United States. However, since the congestion problem facing the Northern Virginia region is substantially more severe than in most other regions, creating and sustaining an effective IM program and linking IM with ITS deployment are probably more important. In recognition of the role that accurate benefit estimates can play in creating and sustaining IM efforts and the application of ITS technology in IM, the interest in benefit estimation approaches has increased. The research discussed here was conducted for the Virginia Department of Transportation (VDOT). It had two distinct but related purposes:

- to develop and apply an approach for estimating the benefits that can result from various ITS applications to IM; and
- to generate estimates of the benefits of several ITS enhanced IM scenarios for the Northern Virginia region.

A combination of library and field research, interviews, and simulation modeling was employed in the approach developed for estimating the benefits of IM. This approach builds on recent efforts at incident prediction and benefit-cost analyses related to alternative IM strategies. It was intended to highlight a means by which agencies can better describe and estimate the benefits of current IM efforts as well as ITS enhanced IM scenarios on a regional basis. Estimated delay reduction benefits of various IM and ITS-enhanced IM scenarios, and a preliminary valuation of the delay reduction benefits were generated for primary roadways in Northern Virginia. These represent a substantial portion of the road network, traffic and congestion in the National Capital Area.

BENEFIT ESTIMATION APPROACH

A review of the literature on IM, ITS and benefit estimation was conducted to identify state-of-the-art approaches to benefit estimation, suitable for "what if?" analyses with ITS-augmented IM scenarios. Information was obtained from a variety of sources, including ITS America, the Texas Transportation Institute, and PATH databases. In parallel, a series of field site visits and interviews with local authorities responsible for or involved in IM was conducted to identify general ITS technology areas and specific applications that were expected to significantly assist local agencies in achieving regional IM benefits. Interviews were conducted with VDOT staff of the Northern Virginia District Office and the Traffic Operations Center in Fairfax, VA, the Traffic Management Center in Arlington, VA, and the Intelligent Transportation Systems Office in Richmond, VA. Also interviewed were staff of the University Center for Transportation Research, Virginia Polytechnic Institute and State University in Blacksburg, VA, the Traffic and Parking Services Division of the Montgomery County Department of Transportation in Rockville, MD, and Maryland State Highway Administration field offices (Padgett, 1996).

Incident Impact Assessment Model

Upon reviewing the relevant literature and considering various alternatives, including the development of a model based on an analysis of existing IM and ITS cost-benefit studies, the IMPACT v1.0 incident impact modeling software was selected as the most promising benefit estimation tool for the Northern Virginia application. IMPACT v1.0 is a computer model of incident occurrence, location and severity that is intended for use in estimating incident impacts for urban freeway segments and traffic volumes, in terms of resulting delays. In addition, it can be used for the quantification of expected changes in these delays, corresponding to alternative improvements to freeways, traffic management, and incident management procedures. Designed with planning and cost-benefit analysis applications in mind, IMPACT v1.0 is a standalone computer program implemented in the Windows environment with standard Windows file management and help screen facilities. The application is entirely menu driven through its stages of input data specification, solution calculation, and output of results (Sullivan and Champion, 1995). The software was developed under contract DTFH61-93-C-00015 between the FHWA, US DOT, and Ball Systems Engineering, with the latter subcontracting to the California State Polytechnic University at San Luis Obispo.

Based on the best available incident, highway geometry and traffic volume data from eight major US cities, the model comprises four sets of calculations.[1] These four sub-models link together to estimate the impact of each of seven categories of freeway incidents,[2] under any one of five IM scenarios.[3] Calculations are performed on a road section by road section basis, with no spill-over effects from one section to the next.

First, the number of incidents of each type is estimated in the incident rate sub-model. This number is then multiplied by percentages associated with various lateral locations of incidents on a roadway section obtained from the location and severity sub-model. In the next step, the number of incidents for each lateral location, together with its characteristic reduction of capacity, is combined with four percentile values that represent the duration distribution of incidents of a given type, occurring in one of two lateral locations, under a given IM scenario. Finally, each combination of capacity reduction and incident duration is evaluated in the delay sub-model and the estimated annual number of hours of delays resulting from each type of incident under a given

IM scenario is generated. A summary total of these delay hours is provided for each road section, and for all road sections in the analysis. Assessing the delay reduction effects of various IM scenarios is a three-step process. First, the "no incident management" scenario implemented in the modeling software must be run to establish baseline values for delays on each road section.

Next, the user must either run one of the three IM scenarios that are implemented in the software or a "user defined" IM scenario. Finally, in order to arrive at the estimated delay reduction due to a particular IM strategy, the delay estimates for that IM scenario must be subtracted from those under the "no incident management" scenario. As implemented, the software includes default values for most model parameters. Consequently, analyses can be performed even if knowledge of specific local values is lacking for a majority of model parameters (Sullivan *et al.*, 1995). Local road section geometry and traffic volumes constitute the minimum required data inputs. Comparison of local accident data and results from trial runs with the IMPACT model suggests that reasonably accurate estimates of the number of incidents (Behnke, 1996) can be obtained with these default values. Therefore, extensive local calibration of the model was not necessary. The general set-up and method of use of the model was suitable to the estimation problem at hand, with two exceptions. First, the model allows the user to select only a single IM strategy while the Northern Virginia IM effort employs all three in combination. Second, no ITS-enhanced IM scenarios are provided in the model.

Model Adaptation

In order to use the model for the specific purposes of this research, two adaptations were made. The first was the implementation of a baseline scenario reflecting the existing IM effort in Northern Virginia, and the second was the development of several ITS-enhanced IM scenarios and their implementation in the model. It was possible to implement the required adaptations using the "User Defined" option in the software's "Incident Management Strategy" dialog box that is part of the delay sub-model. A "User Defined" IM scenario may be set up by providing new percentile values for each of seven incident types. When the user does not specify percentile values, the "User Defined" option defaults to settings that are identical to those under the "No Incident Management" option. Following procedures described in the original software development report, the percentile values of an incident duration distribution can be derived from mean and standard deviation values of incident duration (Sullivan *et al.*, 1995).

In the absence of valid and reliable local incident duration data, a baseline for the IM effort existing in Northern Virginia was constructed by selecting percentile values from among the ones already implemented in the model.

That is, the set of model parameters that yielded the greatest reduction in incident duration for each type of incident was selected. The underlying rationale was that if the three IM strategies were employed concurrently, their delay-reducing effects would be cumulative although not additive. It was assumed that this "optimal" selection from the available incident duration distribution parameters could adequately reflect the combination of traffic management centers, freeway service patrols, and major incident response teams employed in Northern Virginia. In the worst case, underestimation of the delay-reducing effect of the scenario for the combination of all three IM strategies would occur and is, therefore, a conservative approach.

Underlying the incident duration distribution percentile values of the newly specified "Northern Virginia Baseline" IM scenario are mean incident duration values and associated standard deviations for each type of incident, by lateral location of the incidents. Using these means and standard deviations, seven "percentile value look-up tables" were created, one for each type of incident. These tables show the mean incident duration – starting with the baseline mean and followed by means at one minute decreased increments – and the adjusted standard deviations, which are proportional to the means. In addition, they show the incident duration distribution percentile values associated with each set of one mean and one standard deviation. These tables were subsequently used to assess the ITS enhanced scenarios that were developed in parallel with the model adaptations (Baird, 1996).

Scenario Development

To gather the information necessary for the development of new IM scenarios, a literature review and field site visits were conducted. The information obtained suggested that four main ITS technology areas hold promise of real-world benefits in the Northern Virginia region:

- advanced surveillance and detection applications;
- response vehicle communications integration;
- routing and scheduling algorithms for response vehicles; and
- location applications for response vehicles.

On the basis of expert opinion expressed in interviews, four specific ITS applications were selected. These are expected to yield the greatest savings in incident detection, verification, response and/or clearance time and thus have a substantial impact on the IM process and subsequent delay reduction. All four applications have a short- to medium-term field deployment horizon.

To be able to model the effects of the deployment of each of these applications, it was also necessary to obtain estimates of the time savings that each

ITS application could generate. Since data illustrating the time saving effect of ITS applications in IM are virtually non-existent, interviews with IM program officials proved the only viable alternative source of information.

Hence, in addition to general information, local IM experts were asked to provide an indication of the impact they expected from deployment of each of the four ITS applications. Specifically, they were asked to provide an estimate of the reduction of incident duration for all incidents. The four ITS applications and their associated minimum and maximum average incident duration reduction effects are:

- closed circuit TV [CCTV] (4–7 minutes);
- cellular phone in response vehicles (2–5 minutes);
- computer-aided dispatch screens in response vehicles (2–5 minutes);
- global positioning system location for response vehicles (4–7 minutes).

The reduction of the average incident duration for each of the four ITS applications was given under the assumption of full deployment of single applications in Northern Virginia. However, deployment of two or more of the applications simultaneously is likely, and is expected to result in some overlapping of incident duration-reducing effects. As with combinations of IM strategies, the effects of concurrent deployment are considered cumulative but not necessarily additive. Listed below are the five ITS-enhanced IM scenarios that were examined with the IMPACT modeling software, each with applicable values for the reduction of average incident duration:

- CCTV only (4 minutes);
- CCTV and cellular phone (6 minutes);
- CCTV, cellular phone, and GPS (9 minutes);
- CCTV, cellular phone, GPS, and CAD (minimum effect) (13 minutes) (Jin and Stough, 1996);
- CCTV, cellular phone, GPS, and CAD (maximum effect) (19 minutes).

Having determined how to adapt the model to the estimation problem at hand, the modeling of the Northern Virginia Baseline Scenario and the five ITS-enhanced incident management scenarios was accomplished with relative ease. Only the task of input specification remained before estimates of the delay-reduction benefits of ITS deployment in IM could be generated for Northern Virginia.

Input Specification

Input specification required, first, the determination of the road sections to be included in the estimation procedure. The research focused on nine "roadway archetypes" which represent typical situations that may be found in metropolitan areas. An example of each archetype was identified for the Northern Virginia region. Together, these examples provided complete coverage of the Northern Virginia portion of the Capital Beltway (I-495), plus the two main interstate highway routes that feed into Washington, DC from Virginia (I-95 and I-66). The portions of the feeder routes inside and outside the main circumferential (Capital Beltway) were treated as distinct. Portions of two arterial routes (Rt-28 and Rt-123) were included and may be considered secondary circumferentials further away from the main urban concentration. A portion of US-50 was included to serve as a basis of comparison with I-66 and/or I-95 outside the Beltway: all three routes function as spokes to the regional hub, Washington, DC.

Next, the necessary data elements for each roadway segment were collected and/or constructed. The roadway segments were sub-divided into a total of 36 smaller roadway sections, based on the heterogeneity of the roadways as evident from VDOT highway performance data definitions. For each roadway section, the first data items specified were *mileage and average annual daily traffic* (AADT). The required highway geometry and performance data were obtained from VDOT (Commonwealth of Virginia, 1997). Data on the *number of lanes and presence or absence of shoulders on roadways* were field verified for all road sections in the analysis. With respect to traffic and road conditions within the roadway sections, five assumptions were made in order to achieve complete specification of all roadway section data elements as required by IMPACT. These assumptions are described below:

1. Travel peaks in the Northern Virginia region suggest that the morning peak period starts between 6 a.m. and 7 a.m. and ends between 9 a.m. and 10 a.m. (3 hours). The afternoon peak period starts around 2 p.m. and lasts until about 7 p.m. (5 hours). Since the average length of trip on the arterial routes is generally shorter than that on the interstate highways, the length of the total peak period on the arterial routes is assumed to be 1 hour shorter. The *daily peak period length* was set at 8 hours total for all interstate highway mileage and at 7 hours total for arterial mileage.
2. Values for the *peak hour share* of ADT ("K-factor" in the IMPACT model) were derived on the basis of available data for the Woodrow Wilson Bridge on the National Capital Beltway, I-95/495, and validated in trial runs with the model. The K-factor value was set at 8 per-

cent for Interstate highways. For arterial routes, a slightly lower value of 7 percent was used, again in view of the shorter (Stough *et al.*, 1994) average lengths of trips on such roads.

3. It was assumed that, during peak hours, traffic on circumferential routes (such as I-495, Rt-28) has no heavily dominant direction. For roadway sections that are part of a circumferential, the peak hour directional factor D was set at 50 percent. For roadway sections that are part of through routes or commuter routes, however, the directional factor D has initially been set at 70 percent to reflect that in the morning most traffic is "in-bound" and in the evening most traffic is "out-bound" relative to the national capital. For some sections in the analysis, the values for the D factor were subsequently adjusted downwards because high D values caused the IMPACT model's limit to the maximum throughput of sections to be exceeded. This limit is based on the assumed average speed and the recommended throughput design of roadway sections.[4]

4. Depending on the truck definition employed, about 12–15 percent of the AADT generally consists of trucks. During peak periods, the percentage trucks drops to about 5–7 percent. Truck drivers tend to avoid congested regions such as Northern Virginia during peak periods of the day, if they can. In view of this, *the percentage of trucks during peak periods* was set to 0 percent on truck embargo routes (for example, I-66 inside the Beltway), 4 percent on commuter routes, and 6 percent on truck through routes (such as I-95).

5. The value for the *ratio of average annual weekday traffic to daily traffic* (AAWT/AADT) was uniformly set at 1.15. This measure of weekday intensity of traffic for this region was based on a downward adjustment of the actual ratio at the Woodrow Wilson Bridge (I-95/495), which is 1.164. This ratio (1.15) may be interpreted to mean that the region experiences 15 percent more vehicles on an average workday (that is, Monday through Friday) than on an average weekday (that is, Monday through Sunday).

Finally, the data for the 36 stretches of roadways were entered into the IMPACT model, and the total yearly delay for all these sections calculated under the "no incident management" scenario. The required parameters for the "User Defined" IM scenarios had already been derived and the model's incident management type parameters could be varied accordingly. The model was run with the six sets of incident duration distribution parameter values, reflecting the Northern Virginia Baseline scenario and the five ITS-enhanced IM scenarios. The results are summarized and discussed in the next section.

RESULTS

Using identical road section input data for each scenario, the IMPACT model was run seven times with each representing a different IM scenario. Total annual vehicle-miles of travel (VMT) for the 36 road sections in Northern Virginia was estimated at 2 900 000 000, with 55 percent of VMT occurring during peak travel hours. The total number of incidents on these roadway segments (accidents and vehicle fires, mechanical and electrical breakdowns, dropped loads and debris on the road, vehicle stalls, flat tires, and abandoned vehicles) was estimated to be about 34 000 annually, with 77 percent of the incidents occurring during peak hours. Of this, the total number of accidents and vehicle fires was estimated to be 3200, or about 9 percent of the all incidents on the 36 sections. These figures remained constant throughout all seven model runs. That is, only the estimates for the delays resulting from these incidents varied.

Without IM, these 34 000 incidents would have resulted in an estimated 4 242 000 hours of delay annually. In monetary terms, with one hour of delay uniformly valued at $10, this is roughly equivalent to a yearly recurring loss to the region of $42 million. Incident-related delays dropped to 2 786 000 hours per year under the scenario representing the existing IM effort in Northern Virginia, that is, a combination of TMCs, MIRTs and FSPs. In other words, the current IM effort has already reduced delays on the 36 road segments by 1 456 000 hours annually (34 percent). This delay reduction may be conservatively valued at $14.5 million per year. The delays are most likely costlier than this each year, if all interstate and arterial road segments are added in, and higher time valuation is included for trucks and freight, for example. The assignment of a $10 time value yields a conservative estimate of monetary savings as heavy truck delay costs for driver, vehicle, and freight fall anywhere in the range from $25 to $100 per hour. Thus, given 5–7 percent trucks on the roadways, actual savings may be greater than reported here. Accidents and vehicle fires as a share (9 percent) of total incidents conform to the national average (Sullivan *et al.*, 1995). For the Northern Virginia region, however, it is possible that the model underestimates the share of accidents and vehicle fires. This particular region has one of the highest per capita income levels in the nation and, correspondingly, one of the newest and most expensive automobile fleets. With a fleet of this type, one would therefore expect fewer mechanical breakdowns, flat tires, stalls, and abandoned vehicles than in other regions. If, in reality, accidents and vehicle fires were to constitute a larger proportion of total incidents then one would expect more congestion and more delay. Since the 9 percent share seems low for the region, the number of hours of delay presented here may be conservative. In view of the relatively simple method of valuation that was applied, the delay cost estimates are most certainly conservative.

Five specific ITS-enhanced IM scenarios were developed. Since concurrent deployment of all four ITS technologies and subsequent realization of their maximum estimated benefits (ITS-enhanced IM scenarios 4 and 5) may be somewhat optimistic in the short to medium run, only the results of the first three scenarios are discussed. These scenarios appear the most realistic ones within the near future time horizon. The results for the first three ITS-enhanced IM scenarios were compared with those of both the "no incident management" scenario and the scenario reflecting the existing IM capability in Northern Virginia, and estimates of additional benefits derived from (combinations of) ITS applications.

With the current IM effort augmented by full deployment of closed circuit television (scenario 1), annual hours of delay on the selected sections in Northern Virginia dropped further to 2 192 000. As compared to no IM at all, this is a reduction of 48 percent. In this scenario, the ITS augmentation results in a 21 percent gain over the IM effort that existed in Northern Virginia at the time of this study. The additional benefits from deploying CCTV for IM is worth about $6 million per year and would raise the total annual savings as a result of IM in Northern Virginia on the selected segments to $20.5 million (Lo *et al.*, 1994).

Using a combination of CCTV and cellular phones in response vehicles to augment current IM efforts (scenario 2), annual delays on the roadway sections dropped to 1 797 000 hours. That represents a reduction by 58 percent relative to the scenario with no IM activity whatsoever, and an improvement of 35 percent over the region's current IM scenario. The second ITS-enhanced IM scenario is thus worth about $10 million per year, and raises the total annual savings due to incident management to as much as $24.5 million for the Northern Virginia road sections that were analyzed.

In the third ITS-enhanced IM scenario, which adds in-vehicle GPS location to CCTV and cellular phone-based communications, the annual hours of delay are reduced even further to 1 515 000. Compared to having no IM capability in place, this is a total delay reduction of 64 percent. The delay reduction over the existing IM effort in Northern Virginia would be 46 percent. The total annual savings due to IM increases to $27.3 million for the selected road segments; the additional benefits from the third ITS-enhancement scenario is worth $12.5 million in savings. The delay estimates for the Northern Virginia region are considered as accurate as they can be, considering the present implementation of the IMPACT model and the available data. However, some features of both the model and the region require careful consideration when interpreting the results. The absence of spillover effects from one road section to the next has already been mentioned. In addition, it is noted that the model calculates highway segment capacity on the basis of procedures and default values taken from the *Highway Capacity Manual* (Federal Highway Administration, 1985). These values can be modified, but only over a limited range. The calculated section capacities "have considerable influence on the subsequent calculations of incident delays" according to the manual for the modeling software (Sullivan and Champion, 1995). These procedures and default values may play a role in underestimating highway segment capacity as it exists in the Northern Virginia region:

1. In this region, actual highway segment capacities may be greater than assumed in the model. This is due to several factors that have become increasingly characteristic of the management of the region's highway network (Mobility 2000, 1990a). These factors include opening shoulder lanes for travel during peak hours, opening dedicated express lanes for travel during peak hours, and the use of reversible HOV lanes.
2. Actual traffic volumes in this region are significantly in excess of model assumptions; weekday peak volumes on Interstates are typically 150–200 percent of design capacity (Greater Washington Board of Trade, March 1997c). Local transportation officials and IM experts believe that the region has been able to achieve these extraordinarily high volumes through a combination of high rush hour speeds, dangerously close vehicle spacing, and round-the-clock driving.
3. This region exhibits significant "peak spreading" when compared to other regions. That is, traffic volume builds earlier and continues longer in both the morning and afternoon peak periods, making the peak periods longer while enabling these higher traffic volumes to move through the highway network (Mobility 2000, 1990).

In view of the freeway capacity-enhancing features that are employed in this region, and the subsequently greater-than-assumed traffic volumes, the model in its current form probably underestimates true highway capacity. Hence, the traffic volumes accepted by the model, the estimated number of incidents, and the resulting incident delays presented here are likely to be less than they are in reality. The true delay and delay cost estimates in Northern Virginia are therefore probably greater than the estimates developed for the roadway sections in the analysis.

Last but not least, it should be noted that the calculations are performed on the basis of traffic volumes and highway geometry in a given year. Strictly speaking, the estimates that are generated by the model therefore apply to that year only. Highway geometry, however, is unlikely to vary greatly from one year to the next. Further, while annual changes in traffic volumes may be substantially greater than those in highway geometry, they are not expected to be greater to such an extent as to negate the usefulness of the model in analyzing scenarios with a short- or medium-term planning horizon. As the planning horizon is expanded, however, dynamic models are likely to be more appropriate. A more thorough assessment of the sensitivity of the model to annual changes in traffic volumes and highway geometry is in order. However, since this would require extensive testing against historical data, that task was deferred for future research.

CONCLUSIONS

Although the IMPACT model has some shortcomings, it is based on some of the best available data from significant sources of experience with IM in the country. Many assumptions were build into the model by its developers, and many assumptions were added in this research due to lack of precise information. However, as the developers of IMPACT have realized, there are valid analytical procedures that can be employed sensibly. It is concluded, therefore, that while the IMPACT model can be upgraded in a variety of ways, it is among the best currently available. Furthermore, the IMPACT model can be used to provide benefit estimations not only of various IM strategies independently but – with some recalibration and adaptation – also of combinations of IM strategies and ITS-enhanced IM strategies. It can play a significant role in planning, justifying, and funding continued incident management efforts and the deployment of ITS in incident management. With respect to the estimation of the benefits of ITS applications in IM, this research illustrates that ITS technology provides the greatest impact in the early stages of the incident management process, and has an immense effect on the yearly hours of delay saved.

In Northern Virginia, an estimated 35 percent reduction of incident delays has been realized with the existing IM strategies and systems. It estimated that an additional 21–46 percent delay reduction is possible with full deployment of selected applications of ITS technology on roadways, traffic management centers, and in response vehicles. In other words, the effectiveness of the existing incident management efforts may be doubled.

ACKNOWLEDGMENTS

The research project "Benefits Evaluation of Incident Response and ITS: Accidents and Non-recurrent Congestion in the National Capital Region" was conducted with sponsorship from the Virginia Transportation Research Council (VTRC) for the Virginia Department of Transportation (VDOT). The authors wish to acknowledge the support of both VDOT and the VTRC in making this project possible. This research took place between March 15,

1996 and June 30, 1997, and the project was completed with the delivery of the report entitled *Final Report: Application of the IMPACT Model to ITS Benefit Estimation.* Copies of this report can be obtained by writing to The Institute of Public Policy, George Mason University, 4400 University Drive (MS. 3C6), Fairfax, VA 22030-4444, or by calling 703-993-2268.

NOTES

1. The four sets of calculations respectively deal with (a) freeway capacity, (b) incident rates, (c) lateral distribution of incidents and resulting capacity reduction, and (d) incident duration and resulting delays.
2. The seven categories of incidents distinguished in the model are: (a) accidents and vehicle fires, (b) major mechanical and electrical breakdowns, (c) dropped loads and other debris on the highway, (d) vehicle stalls, (e) flat tires, (f) abandoned vehicles, and (g) all other.
3. The five IM scenarios distinguished in the model are: (a) no incident management, (b) traffic management center (TMC), (c) major incident response team (MIRT), (d) freeway service patrol (FSP), and (e) user defined.
4. The National Capital region is among the most congested in the country. Yet the actual through-put on road sections often exceeds throughput design values. The model accommodates such situations by accepting throughput values of up to 15 percent greater than design throughput, without this resulting in any error in the model's calculations. However, for some road sections, the actual throughput exceeds design throughput by more than 15 percent. In these cases adjustments must be made in order to avoid error messages and overflow in the results. The IMPACT model allows, although with maximum caution, for the default design throughput values to be changed. However, when changes are made, the model appears to become un-stable, producing results that vary to great extremes. Therefore, rather than tampering with these defaults at the risk of model instability and error, minor changes were made in the inputs instead.

REFERENCES

Baird, J.K. (1996), "IHVS and transportation demand management: meeting the chal-lenges together?", *IHVS Policy Workshop in Institutional and Environmental Issue*, Monterey CA: Asilomar Conference Center, pp. 81–95.

Behnke, R.W. (1996), "Advanced Public Transportation Systems (APTS): multimodal and alternative market applications of IVHS", *IVHS Policy Workshop in Institu-tional and Environmental Issue*, Monterey CA: Asilomar Conference Center, pp. 19–36.

Cambridge Systematics Inc. (1990), *Incident Management. Final Report Prepared for Trucking Research Institute*, Alexandria, VA: ATA Foundation, Trucking Research Institute.

Cambridge Systematics Inc. (1997), *Incident Management: Challenges, Strategies, and Solutions for Advancing Safety and Roadway Efficiency* (Full Report), Alexan-dria, VA: ATA Foundation, Trucking Research Institute.

Commonwealth of Virginia (1997), *Average Daily Traffic Volumes on Interstate, Arterial and Primary Routes. 1995*, Richmond, VA: Virginia Department of Trans-portation.

Federal Highway Administration (1985), *Highway Capacity Manual, 1985*, Washington, DC: FWHA, US Department of Transportation.

Greater Washington Board of Trade (1997a), *Transportation Study 1997. Report 1: Historical and Current Conditions*, Washington, DC: Greater Washington Board of Trade, pp. 6–7.

Greater Washington Board of Trade (1997b), *Transportation Study 1997. Report 3: Economic and Quality of Life Costs of Not Meeting Transportation Needs Conditions*, Washington, DC: Greater Washington Board of Trade, pp. 4–5, 7.

Greater Washington Board of Trade (1997c), *Transportation Study 1997. Report 2: Regional Forecasts and Constrained Long Range Plan Assessment*, Washington, DC: Greater Washington Board of Trade, p. 6.

Greater Washington Transportation Planning Board (1990), *The Highway and Transit Facility Element of the Long-Range Transportation Plan for the National Capital Region*, Washington, DC: Greater Washington Transportation Planning Board, December.

Haynes, K.E., Phillips, F.Y., Qiangsheng, L., Pandit, N. and Arieira, C.R. (1994), "Information based uncertainty management of IVHS/IS-T technologies", paper presented at the International Symposium on IVHS/GIS-T in Seoul, South Korea.

Institute of Transportation Engineers (1996), *Mobility Facts*, Washington, DC: Institute of Transportation Engineers.

Jin, D.J. and Stough, R.R. (1996), "Paradigms and ITS evaluation and deployment", paper submitted to the Transportation Research Board for presentation at the 1997 Annual Conference the Institute of Public Policy, Fairfax, VA: Institute of Public Policy, George Mason University.

Lo, H.K., Chatterjee, A., Wegmann, F., Roberts, S. and Rathi, A.K. (1994), "Evaluation framework for IVHS", *Journal of Transportation Engineering* **120**(3), 447–60.

Mannering, F.L., Hallenbeck, M. and Koehne, J. (1992), *A Framework for Developing Incident Management Systems: A Summary*, report WA-RD 224.2, Seattle: Washington State Transportation Center, University of Washington.

Mobility 2000 (1990a), *Final Report of the Working Group on Operational Benefits*, San Antonio, TX: Mobility 2000, March.

Mobility 2000 (1990b), *Proceedings of a Workshop on Intelligent Vehicle Highway Systems*. San Antonio, TX: Mobility 2000.

Padgett, R.L. (1996), "Human error and risk management approaches to cost-effectiveness and decision making in intelligent transportation system development and evaluation", dissertation submitted for the degree of Doctor of Philosophy in Information Technology, Fairfax, VA: George Mason University.

Stough, R.R., Maggio, M.E. and Jin, D.E. (1994), *Methodological Challenges in Regional Evaluation of ITS Induced and Direct Effects*, Working Paper, Fairfax, VA: The Institute of Public Policy, George Mason University.

Sullivan, E. and Champion, D. (1995), *IMPACT 1.0: A Model for the Estimation of Incident Impacts on Freeways. User's Guide*, Washington, DC: FWHA, US Department of Transportation.

Sullivan, E., Taff, S. and Daly, J. (1995), *Final Report for Tasks H–J Incident Detection Issues: a Methodology for Measurement and Reporting of Incidents and the Prediction of Incident Impacts on Freeways*, report DTFH61-93-C-00015, Washington, DC: FWHA, US Department of Transportation.

US Department of Transportation, Federal Highway Administration (1987), *Highway Statistics 1987*, Washington, DC: US Government Printing Office.

Index